T0260415

Blockchain

Blockchain: Principles and Applications in IoT covers all the aspects of blockchain and its application in IoT. The book focuses on blockchain, its features, and the core technologies that are used to build the blockchain network. The gradual flow of chapters traces the history of blockchain from cryptocurrencies to blockchain technology platforms and applications that are adopted by mainstream financial and industrial domains worldwide due to their ease of use, increased security, and transparency.

- Focuses on application of blockchain on IoT domain
- Focuses on blockchain as a data repository
- Goes beyond blockchain in relation to cryptocurrency to cover its use in areas like health care, supply chain management, etc.
- Covers consensus algorithms like PAROX, RAFT, etc. and their applications

This book is primarily aimed at graduates and researchers in computer science and IT.

Blockchain
Principles and Applications in IoT

Edited by
Rajdeep Chakraborty, Anupam Ghosh,
Valentina Emilia Balas, and Ahmed A Elngar

CRC Press
Taylor & Francis Group
Boca Raton London New York

CRC Press is an imprint of the
Taylor & Francis Group, an **informa** business

A CHAPMAN & HALL BOOK

Front cover image: Immersion Imagery/Shutterstock

First edition published 2023
by CRC Press
6000 Broken Sound Parkway NW, Suite 300, Boca Raton, FL 33487-2742

and by CRC Press
4 Park Square, Milton Park, Abingdon, Oxon, OX14 4RN

CRC Press is an imprint of Taylor & Francis Group, LLC

Library of Congress Cataloguing-in-Publication Data
Names: Chakraborty, Rajdeep, editor.
Title: Blockchain : principles and applications in IoT / edited by Dr. Rajdeep Chakraborty, Dr. Anupam Ghosh, Professor. Valentina Emilia Balas, Dr. Ahmed A. Elngar.
Other titles: Blockchain (CRC Press)
Description: First edition. | Boca Raton : Chapman & Hall/CRC Press, 2023. | Includes bibliographical references and index. |
Identifiers: LCCN 2022015553 (print) | LCCN 2022015554 (ebook) | ISBN 9781032068060 (hbk) | ISBN 9781032068091 (pbk) | ISBN 9781003203957 (ebk)
Subjects: LCSH: Blockchains (Databases)
Classification: LCC QA76.9.B56 B48 2023 (print) | LCC QA76.9.B56 (ebook) | DDC 005.74--dc23/eng/20220527
LC record available at https://lccn.loc.gov/2022015553
LC ebook record available at https://lccn.loc.gov/2022015554

ISBN: 978-1-032-06806-0 (hbk)
ISBN: 978-1-032-06809-1 (pbk)
ISBN: 978-1-003-20395-7 (ebk)

DOI: 10.1201/9781003203957

Typeset in Palatino
by MPS Limited, Dehradun

Contents

Section C Security Aspects in Blockchain and IoT

Preface

Blockchain is a distributed ledger system to keep all the data in a chain of blocks. This system is basically a server less system and the node works in mutual consensus. The basic principle is, first a transaction/data is created and it is agreed by n/2+1 nodes of 'n' number of nodes, then hash is generated to this new data plus previous block, if exist, and finally data+hash is the block generated to be added in the blockchain. Once a data is added then no modification or deletion is possible in future. Principally, blockchains are of mainly two types, private/permission blockchain and public/permission-less blockchain. Bitcoin and Ether are famous blockchains.

Blockchains have many applications in the IoT domain. Blockchain can be used for traffic data system, sensor data system, supply chain management and one important application is no-code or low-code software development. Blockchain technology is the answer to some of the primary challenges of IoT, including its scalability, privacy, and reliability. Blockchain technology can be an excellent means to track and monitor billions of connected devices, thereby enabling the sharing and processing of transactions between connected devices. Moreover, being decentralized, it would eliminate single points of failure, creating a more resilient ecosystem for devices to run on. Blockchain IoT can facilitate secure and reliable collaboration between connected devices in an IoT network.

Thus, this book will cover all the aspects of blockchain, its application in IoT, and all the security aspects. Though this book is edited, it is designed such that it can be used as a textbook/reference book in UG and PG courses having blockchain as a subject.

This book is organized into three sections, "Section A: Blockchain Principles" describes the overview of blockchain and IoT in Chapter 1. Next, all the important consensus algorithms are covered in Chapter 2. Smart contracts, as one of the important features through which blockchain offers services to users, is covered in Chapter 3. Chapter 4 covers the Internet of Things (IoT) in detail. The implementation challenges in blockchain are covered in Chapter 5. Finally, Chapter 6 covers how blockchain and IoT can be a benchmark in Industry 5.0. "Section B: Blockchain Applications in Internet of Things (IoT)" describes ASIC-based mining system in blockchain for IoT environment and also discusses scalability and security issues in Chapter 7. Chapter 8 shows how blockchain can be used as patient record management and analysis using machine learning approach. Chapter 9 is about an IoT EHM system using blockchain. Chapter 10 tells us why it is important to study blockchain and IoT as university courses (UG, PG, and PhD) in the context of India and internationally. The final section, "Section C: Security Aspects in Blockchain and IoT," gives a detailed understanding cryptographic and consensus techniques supporting privacy and security management of cryptocurrency transactions in Chapter 11. As blockchain is itself data repository, so, Chapter 12 covers secured blockchain-based databases for storing sensitive information for access control IoT systems. Chapter 13 gives the necessary cryptographic foundations required to implement a blockchain securely. Chapter 14 covers access control and data security of IoT applications using blockchain technology and finally, Chapter 15 discusses lightweight cryptography and protocols that are nowadays used extensively for blockchain and IoT security.

Section A Blockchain Principles

Chapter 1 gives the overview of blockchain and IoT. It starts with the discussion of various components in IoT. Then this chapter covers a detailed literature review with challenges in IoT in the next section, including security requirements in IoT. Thereafter, this chapter gives the definition of blockchain and types of blockchain. Then it moves towards the component of blockchain with privacy and security requirements of blockchain. Finally, this chapter shows how blockchain is integrated with IoT for various applications.

Chapter 2 covers all the necessary consensus mechanisms. A consensus mechanism is an algorithm that approves transactions or records on a decentralized ledger while rejecting false or fraudulent records. This chapter covers mostly used consensus mechanisms like proof of work, proof of stake, delegated proof of stake, proof of capacity, proof of elapsed time, proof of authority, and proof of activity with their block diagrams, explanation, pros, and cons. This chapter concludes with a relative study of these consensus mechanism.

Chapter 3 covers smart contracts that provide services to the users of a blockchain. It starts with the definitions of the smart contracts and off-chain contracts. Next, it covers the literature followed by workings of the smart contracts with various example platforms. Then it discusses issues related to the security in smart contracts. Another aspect is covering various applications of smart contracts with various use cases. This chapter ends with the research challenges in smart contracts.

Chapter 4 covers the Internet of Things (IoT) in detail. This chapter starts with the introduction of IoT and its various applications. Then it covers the generation of IoT life cycle followed by amalgamation of blockchain for IoT applications. Thereafter, it discuss network slicing of IoT and IoT-enabled technologies. This chapter ends with research challenges and the future scope of IoT in our world.

Chapter 5 gives details about blockchain implementation challenges in IoT. It discusses the research challenges in blockchain like, scalability, integrated cost, privacy and security challenges, regulation issues, energy consumption, lack of standardization, selfish mining, lack of understanding, and awareness with a detailed literature coverage. Next, this chapter gives mathematical foundations of blockchain with the difference between blockchain and mining applications. Some of the cryptographic foundations are also discussed in the last part.

Chapter 6 covers how blockchain and IoT can be a benchmark in Industry 5.0. It starts with IoT overview with benefits, drawbacks, and applications. Then the chapter moves towards IoT security. Thereafter, this chapter discuss the workings of IoT. It concludes with various smart applications of IoT in Industry 5.0.

Section B Blockchain Applications in Internet of Things (IoT)

Chapter 7 gives an implementation of ASIC-based mining system in blockchain for IoT environment to increase scalability and security. This chapter starts with the present problems in IoT applications using blockchain and gives the proposed solution. Then it gives the detailed understanding of blockchain in implementation perspective. In the next section, it gives the proposed model in details. This chapter concludes with ASIC miners and a comparative study of why one should opt for ASIC mining based blockchain for IoT applications.

Chapter 8 shows how blockchain can be used for patient record management and analysis with machine learning. It starts with preliminaries and background of the proposed system. Next, it moves towards importance of health care, followed by machine learning approaches. Finally, this chapter concludes with a future scope.

Chapter 9 gives an IoT-based electronic health records (EHRs) management system using blockchain technology. This chapter starts with the importance and structure of the blockchain followed by blockchain components and actors. Then it gives the workings of blockchain. Then it gives the proposed model and implementation with various types of blockchains. Thereafter it gives the proposed system, working philosophy, and algorithms. It concludes with the importance and overview of the system.

Chapter 10 tells us about why it is important to study blockchain and IoT as university courses (UG, PG, and PhD) in the context of India and internationally. This chapter is all about e-learning, regular learning, and vocational learning of blockchain and IoT. It also covers detailed courses available worldwide for the study of blockchain and IoT. Finally, this chapter concludes with some proposed courses in blockchain and IoT.

Section C Security Aspects in Blockchain and IoT

Chapter 11 covers cryptographic and consensus techniques supporting privacy and security management of cryptocurrency transactions in details. Cryptocurrency is another important application of blockchain and it requires a high degree of privacy and security. This chapter starts with blockchain fundamentals and transaction process in cryptocurrency. Next, it covers privacy and security issues in detail. Cryptographic techniques are the algorithms and protocols that provide privacy and security, and this is discussed in detail thereafter. This chapter concludes with various consensus mechanisms and their securities.

Chapter 12 gives a database approach using blockchain for storing sensitive information that can be used for access control in IoT systems. After a discussion of blockchain, it discusses Ethereum-based blockchain and smart contracts. Next, this chapter moves towards proposed architecture and implementation. A conclusion is drawn in the last section.

Chapter 13 gives all the cryptographic foundations of blockchain technology. In detail, it starts with cryptographic hash function and then discusses public key cryptography. The post quantum cryptography concept is covered next. Then it moves to zero knowledge protocols and lightweight cryptography, which are needed for IoT and blockchain security.

Chapter 14 provides an access control and data security model for IoT applications using blockchain technology. It starts with various access control methods and challenges. Then it gives this scenario with regard to blockchain technology. Then it discusses blockchain-based access control in various domains. This chapter concludes with future directions and research issues.

Chapter 15 discuss the lightweight cryptography and protocols, which are now the only ways to provide security in low resource IoT applications. First, it gives a detailed discussion of lightweight block ciphers like PRESENT, TEA, and CLEFIA, and then it gives a detailed discussion of lightweight stream ciphers like GRAIN V1, MICKEY, and TRIVIUM, and thereafter it gives a detailed discussion of lightweight cryptographic hash functions like PHOTON, SPONGENT, and QUARK, and finally it gives a detailed discussion of lightweight security protocols like MQTT, CoAP, and XMPP. This chapter concludes with a relative discussion of these algorithms in their types.

About the Editors

**Dr. Rajdeep Chakraborty** did his PhD in CSE from the University of Kalyani with M.Tech in IT and BE in CSE. His research experience is of 9 years and academic experience is of 14 years. He has several publications in reputed international journals and conferences. He also has written an authored book on hardware cryptography. His field of interest is mainly in cryptography and computer security. He has also guided many masters' theses and UG projects. He is an enthusiastic and budding researcher in computer security. He has also awarded with Adarsh Vidya Saraswati Rashtriya Puraskar, National Award of Excellence 2019 conferred by Glacier Journal Research Foundation, Global Management Council, Ahmedabad, Gujrat, India and conferred on 10/December/2019. He is also guiding a PhD Scholar under MAKAUT, formerly WBUT, West Bengal, India and nominated secretary of Indo-UK Confederation of Science, Technology and Research, London, UK.

Assistant Professor, Dept. of Computer Science and Engineering
Netaji Subhash Engineering College, Garia, KOLKATA – 700152, India
E-mail: rajdeep_chak@rediffmail.com

**Dr. Anupam Ghosh** did his PhD in engineering from Jadavpur University with M.Tech in computer science and engineering and MSc in computer science from the University of Calcutta. He is presently working as a professor of the Department of Computer Science and Engineering, Netaji Subhash Engineering College, Kolkata. His teaching experience of 19 years includes 13 years in research. He has published more than 80 international papers in reputed international journals (SCI/SCIE/ESCI) and conferences. His field of interest is mainly in AI, machine learning, deep learning, image processing, soft computing, bioinformatics, IoT, and data mining. He has also guided many masters' theses and UG projects.

Professor, Dept. of CSE, Netaji Subhash Engineering College, Kolkata – 700152, West Bengal, India.
E-Mail: anupam.ghosh@rediffmail.com

**Dr. Valentina Emilia Bălaş** at present, is working as Conferenţiar universitar, Facultatea de Inginerie, Universitatea "Aurel Vlaicu" din Arad, Romania. She did her PhD in intelligent electronic and communication and her work domains are system integration, soft computing, fuzzy control, system modeling and simulation, and biometrics. She has completed many international projects and published more than 100 papers in renowned international journals and conferences. She is also editorial board member of many reputed journals such as Neurocomputing, ISSN 0925-2312, ISI impact factor: 0,866, IEEE Transactions on Fuzzy Systems, ISSN 1063-6706, ISI impact factor: 2,262, IEEE

Transactions on Industrial Electronics, ISSN 0278-0046, ISI impact factor: 1,872, IEEE Transactions on Industrial Informatics, ISSN 1551-3203, ISI impact factor: 1,487, IEEE TSMC part B, Transactions on Systems, Man, and Cybernetics Part B, ISSN 1083-4419, ISI impact factor: 2,448, Journal of Intelligent and Fuzzy Systems, ISSN 1064-1246, ISI impact factor: 0,437. She has many awards and recognitions to her name and she is general chair and co-chair of many prestigious international conferences. She has also delivered invited talks, keynote addresses, panel discussions, etc. at various events. She has 25 years of teaching and university experiences. She also worked as a project engineer for 13 years.

Ph.D. and Professor, University of Arad, Faculty of Engineering, Department of Automation and Applied Informatics, 77 B-dul Revolutiei, 310130 Arad, Romania.

E-mail: balas@drbalas.ro; valentina.balas@uav.ro, Mobile: 0040-740-059151

Dr. Ahmed A. Elngar works as an assistant professor, Faculty of Computers & Artificial Intelligence, Computer Science Department, Beni-Suef University, Beni-Suef, Egypt. He is also director of Technological and Informatics Studies Center at Beni-Suef University, Beni-Suef, Egypt; founder and chairman of Scientific Innovation Research Group (SIRG), Beni-Suef University, Beni-Suef, Egypt; deputy director of the International Ranking Office, Beni Suef University; and managing editor in *Journal of CyberSecurity and Information Management* (JCIM). He did his Doctor of Philosophy (Ph.D) of computer science from Faculty of Science, Al-Azhar University – Cairo, Egypt. He has also completed various prestigious certificates and his fields of work are object oriented programming, network security, cryptography and network, information and computer security, fundamental of multimedia, steganography and digital watermark, data warehousing, digital signal processing (DSP), fundamentals of database systems, introduction to algebra and geometry, software engineering, and intelligent systems. He has many renowned publications in international conferences and international journals. He also has various books and book chapters; he has delivered keynote addresses, invited talks, session chairing, and chair and co-chair in various renowned international conferences. He has 10 years of teaching experience and also has 5 years of experience as an instructor and developer.

Assistant Professor, Faculty of Computers and Artificial intelligence, Beni-Suef university, Beni Suef City, 62511, Egypt.

Email: elngar_7@yahoo.co.uk, ahmedelnagar@fcis.bsu.edu.eg, (+2) 01007400752 - (+2) 01147516672

Contributors

A. Ali
Glocal University
Saharanpur, U.P., India

Abhik Banerjee
Netaji Subhash Engineering College
Kolkata, India

Aleem Ali
Department of CSE
Glocal University
Saharanpur, U.P., India

Amit Dua
Birla Institute of Technology and Science
Pilani-Rajasthan, India

Asif Ali Laghari
Sindh Madressatul Islam University
Karachi, Sindh, Pakistan

Awais Khan Jumani
Ilma University
Karachi, Sindh, Pakistan

Barnita Maity
University of Calcutta
Kolkata, India

Bhaskar Dutta
University of Calcutta
Kolkata, India

C. Tezcan
Middle East Technical University
Ankara, Turkey

Debatosh Pal Majumder
Dept. of CSE
Netaji Subhash Engineering College
Kolkata, India

Dr Tun Myat Aung
University of Information
  Technology – UIT
Yangon, Myanmar

Dr. Kamal Kant Sharma
Chandigarh University
Punjab, India

Gunjan Chhabra
University of Petroleum & Energy
  Studies
Dehradun, India

K. Chandrasekaran
Computer Science and Engineering
NIT-Karnataka
Surathkal, Mangalore, Karnataka, India

Md Faisal
Glocal University
Saharanpur, U.P., India

Ni Ni Hla
University of Computer Studies
Yangon – UCSY
Myanmar

P. K. Paul
Executive Director, MCIS, & Assistant
  Professor (IST)
Department of CIS, & Information
  Scientist (Offg.)
Raiganj University
India

R. Kaur
Glocal University
Saharanpur, U.P., India

Rashmi Sharma
University of Petroleum &
    Energy Studies
Dehradun, India

Reshma
Chandigarh University
Punjab, India

Runa Chatterjee
Department of Computer Science and
    Engineering
Netaji Subhash Engineering College
Kolkata, West Bengal, India

Shahid Nazir
Birla Institute of Technology and Science
Pilani-Rajasthan, India

Shaweta Sachdeva
Department of CSE
Glocal University
Saharanpur, U.P., India

Smarta Sangui
Sister Nivedita University
Kolkata, India

Suman Das
Sister Nivedita University
Kolkata, India

Swarup Kr Ghosh
Sister Nivedita University
Kolkata, India

Tamoghna Mandal
National Institute of Technology
Durgapur, West Bengal, India

Trishit Banerjee
Netaji Subhash Engineering College,
    Techno City Garia
Kolkata, India

Usha Divakarla
Information Science and Engineering
NMAMIT, Nitte
Karkala, India

Varun Sapra
University of Petroleum &
    Energy Studies
Dehradun, India

Waqas Ahmed Siddique
Ilma University Karachi
Sindh, Pakistan

Section A

Blockchain Principles

1

An Overview of Blockchain as a Security Tool in the Internet of Things

Shaweta Sachdeva and Aleem Ali

Department of CSE, Glocal University Saharanpur,
U.P., India

CONTENTS

DOI: 10.1201/9781003203957-2

3

1.1 Introduction

IoT points out the interconnection of smart devices that helps to collect data from various devices and make decisions. The Internet of Things (IoT) is transforming the traditional sector into a smart industry based on data-driven decision-making. Decentralization, poor communication, privacy, and safety frangibility are all inherent characteristics of the Internet of Things. Despite this, IoT is vulnerable to secrecy and safety risks due to a lack of intrinsic protective methods [1]. The IoT is vulnerable to secrecy and security flaws due to methodology. The most advanced technology can assist in dealing with IoT's essential safety provisions. By being part of a security framework, blockchain techniques are involved in protecting various IoT-aligned apps. A blockchain is a database in which all processed records or data are reserved in order of processing. Information is saved as a global log that should not be changed. This type of mechanism is required by the IoT to approve secure data sharing between IoT devices in a variety of environments. The blockchain is a cutting-edge way of technology that may help to improve data security while it is being exchanged.

1.2 The Internet of Things (IoT) Definition

As information technology rises, the Internet of Things (IoT) has become a part of our daily routine. It is defined as the process of communication among devices that run on a variety of platforms and have self-contained functions without the need for human intervention. This term arose as a result of the Internet's expansion over the last few decades, portrayed as a new technical and social revolution [2]. It is a subset of the current Internet that allows for data processing and connectivity between linked objects [see Figure 1.1]. Things will be able to be managed remotely and accessible as service providers thanks to their connectivity to the global computer network, making them smart things.

1.2.1 IoT Components

An IoT system has four essential components that explain how it works. All of the components are shown in Figure 1.2.

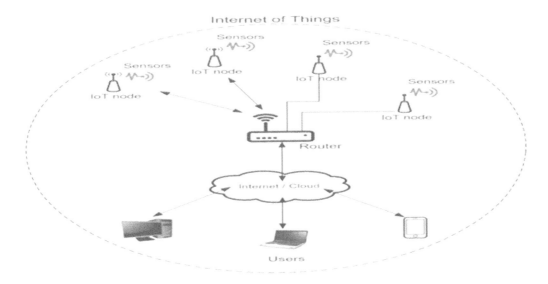

FIGURE 1.1
IoT infrastructure.

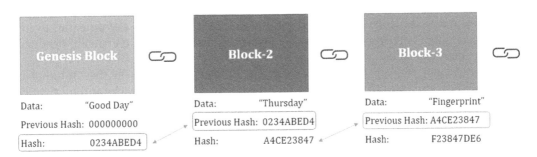

FIGURE 1.2
IoT components.

1.2.1.1 Sensors/Devices

Sensors or gadgets aid in the collection of particularly little data from the environment. Other than sensing objects, many sensors on a device can be stacked together to execute a variety of activities [3]. A phone, for example, has multiple sensors, such as a GPS, accelerometer, and camera, but it also performs other jobs, such as perceiving its surroundings.

1.2.1.2 Connectivity

The data is subsequently delivered to a cloud infrastructure, although a transport medium is required. Sensors can interact with the cloud via cellular networks, satellite networks, Wi-Fi, Bluetooth, and other technologies.

1.2.1.3 Processing of Information

Once the data is obtained, the software processes it before sending it to the cloud.

1.2.1.4 User-Interface Design

After that, the data is made available to the end user in some way. Alarms are set on the users' phones, or SMS or emails are sent to notify them of this.

1.3 Review of the Literature

D. Das et al. proposed a new type of framework for IoT networks that might aid in the speedy creation, development, and deployment of dedicated IoT networks (dubbed "ecosystems") for study or monitoring [4].

K. Lei, et al. proposed the view of Groupchain, which works as a scalable public blockchain with a two-chain structure basically well-suited for fog computing in IoT. Our security research reveals the Groupchain that maintains the security of a Bitcoin-like blockchain while also improving defence against double-spends and selfish mining assaults [5]. It creates a Groupchain prototype and performs tests on it. The results of the experiments show that Groupchain optimizes transaction throughput and confirmation time.

Madumidha et al. proposed the concept of blockchain technology used in early stages of development [6]. BCT has the potential to revolutionize it by streamlining supply chains and, more difficult, offering traceability and security. It has the potential to improve the country's economic condition by lowering corruption and enhancing producer and consumer satisfaction.

Zhicheng Chen et al. proposed an overview of security threats in thee Internet of Things, allowing readers to better grasp the threat posed by these attacks and motivating them to look into creative solutions for the NDN of Things [7]. The essay begins by presenting an overview of the NDN of Things. Security threats on information gathering, caching, and data distribution are investigated. It also goes through the existing security solutions in the NDN of Things.

Travis Mick, et al. proposed design, in order to assist defend the network against routing attacks like blackholes. Before started routing, each node in the network is authentic [8]. We employ symmetric cryptography because asymmetric cryptography is often infeasible on IoT devices.

Yan Zhang et al. proposed the the Internet of Things (IoT) security situation is critical. The IoT has a higher rate of security issues than traditional computer networks. The dispersion and mass of the Internet of Things necessitate dynamic methods to IoT security [9]. This research proposes a revolutionary method to IoT security based on immunology. Traditional network security models are utilised as a guide, and specific IoT security needs are considered. The suggested approach creates a dynamic defense frame for IoT security.

1.4 Challenges in IoT

IoT introduces several services and applications; nevertheless, in order to integrate the virtual and real worlds into a single platform, IoT requires common protocol, methodology, architecture, standards, and security measures. IoT difficulties can be divided into three categories: architecture, standards, and security and privacy concerns.

Every smart device creates data that can be used to provide a wide range of services at any time and in any location. To forecast future decisions, infrastructure is required to support, integrate, and analyse the generated data [10]. As a result, IoT necessitates dynamic architecture in order to introduce a comprehensive blueprint for supporting various objects and applications. In terms of the standard challenge, the IoT standard does not provide enough opportunities for devices to use and access network resources on an equal footing.

Unfortunately, traditional network protocols are insufficient to support growing smart objects and applications. As a result, the IoT network should have a defined standard to enable new objects and applications. As previously indicated, the Internet of Things (IoT) includes a variety of sensors, each of which has its own power limitations. As a result, traditional security measures will fail to protect IoT machine-to-machine connectivity [11]. As a result, IoT requires security procedures in order to obtain low-cost computational power and convenient security solutions.

At successive stages, attackers may use a variety of ways to damage the IoT network. As a result, data security has risen to the top of the priority list in IoT network architecture. As a result, IoT security and privacy issues have become increasingly important in the IoT system.

1.4.1 User Privacy and Data Protection

The Internet of Things (IoT) system is based on transmitting and exchanging information/data among various smart devices via various channels. Users' personal information, such as their personality and behavior, may be included in the data sent [12]. The Internet of Things technology, as previously said, allows any device or person to be automatically identified.

Consumer data can be obtained from their linked objects, which are stored in the system or exchanged across smart objects via various channels. All of a user's sensitive information is subject to the most dangerous assaults and risks if there are no authentication processes in place. As a result, data security and privacy are crucial in the IoT network. Privacy is made up of three main components.

IoT trust management and policy integration are extremely difficult due to the limited protocols, resources, and capabilities of various smart devices. Trust management is required for IoT security, information security, services, apps, and user privacy [13]. Trust management is required for interactions between smart objects for data exchange and administration.

1.4.2 End-to-End Security

There are billions of smart objects, each of which transmits a huge amount of data to other objects. A smart object should be authenticated and have security processes in place to protect people, devices, and services. Simultaneously, security solutions are used to safeguard data and services against threats and attacks. This activity is referred described as "end-to-end security."

A smart object can then connect to the cloud right away. The cloud is in charge of authentication and message control between smart devices [14]. After authentication and control mechanisms have been performed, a smart object connects to the Internet through the gateway. The encryption technology is used to encrypt messages sent between smart devices.

1.4.3 Authentication and Identity Management

Authentication techniques are critical in IoT security, and they can be implemented in a variety of ways, such as ID cards, passwords, and public key infrastructure (PKI). Traditional authentication processes, on the other hand, are useless due to the heterogeneity and complexity of IoT products.

When it comes to identity management, it's utilized to manage smart object IDs, services, and functions. It provides identification, authentication, and access control services. Identity management is used to build smart object relationships. Connectivity, network domains, and applications are all part of the IoT platform. As a result, authentication strength is crucial to identity management [15].

1.4.4 Access Control and Authorization

Authorization and access control methods allow users to get access to network resources and services. Authentication and access control prohibit unauthorized users from accessing network resources. As previously said, IoT devices have limited storage space and power, making authorization and access control approaches difficult to implement. Because of the variety and complexity of devices, authorization and access control approaches may not be suited for IoT systems.

1.4.5 Security Solutions and Threats Resistance

The security management system is used to determine the security processes for the IoT layers. It is used to defend against a wide range of threats and attacks [16]. The security system strategy must include the protection of devices, communication medium, shared data, and services.

The security solutions provide a pleasant working environment for the IoT system, allowing it to adapt to the nature of different devices and applications. IPsec, asymmetric and symmetric encryption, authentication, and other security technologies are available. In a broader sense, it's also required to develop IoT security standards, which are briefly covered below.

1.5 Security Requirements in IoT

The security needs have evolved into a critical component of successful IoT implementation. The concepts of security are applied to the development of security solutions and management systems. The following are the requirements for developing an IoT security system.

1.5.1 Availability

The goal is for consumers to be able to access services at any time and from any location. Maintaining continual communication between users and network resources is crucial. To protect network resources from attacks and threats, all users should be verified. System conflicts and network congestion can restrict data flow, and availability can assist prevent them.

While accountability alone will not prevent IoT attacks and threats, it is vital to maintain and support other security criteria such as integrity and confidentiality [17–20]. It is used to monitor and detect any unknown operations on any device that sends and receives data by setting rules for devices, users, and their actions.

1.5.2 Auditing

For detecting IoT security problems, auditing is a crucial part of the security requirements. It is based on an evaluation and service system. It's used to determine whether or not an IoT system complies with a set of application standards.

1.5.3 Authentication and Authorization

The most critical security need is authentication, which involves using security mechanisms such as public and private keys to identify the user as an authenticated entity (cryptography algorithms). In terms of authorization, it is used to allow users to access network services or resources.

1.5.4 Access Control

Access control is used by network administrators to assign personnel specialized tasks or authenticated access to network resources such as reading, writing, editing, or altering data [21]. As a result, authenticated users can do specified operations thanks to access control.

The term "privacy" relates to the protection of a user's personal information. It has the ability to dissuade illegal users. There are many different levels and types of privacy, such as:

1. Physical and commutation privacy define device privacy. It's possible that information from devices will be stolen.
2. IoT device communications are reliant on privacy throughout communication. It's used to prevent confidential information from being shared during talks.
3. During the processing and storage phases, privacy is applied to protect processed data.
4. Authorized users can see the geographical location of IoT devices thanks to identity and location privacy.

1.5.5 Confidentiality

Confidentiality is a crucial concept in security demands since it prevents unauthorised users from obtaining data [22]. Confidentiality may be used to identify, authenticate, and authorize any smart object in an IoT network. To protect data confidentiality, a variety of security mechanisms, such as authentication protocols, are available.

1.5.6 Integrity

It is a security concept that allows authorised users to access, read, delete, or modify data while adhering to specified guidelines. Internal attacks, which are the most dangerous problem in the network, can be prevented since they are authenticated authorised users. Furthermore, thieves may alter data as it is being transmitted, making it impossible for others to read or edit it [23].

1.5.7 Non-repudiation

The definition of non-repudiation is straightforward. It's used to demonstrate that neither the sender nor the recipient can deny that their messages are their own. Non-repudiation, as a result, can protect against internal attacks.

1.6 Definition of Blockchain

It's a way for untrustworthy participants to come to an agreement. Trustworthy third parties are typically entities such as banks or notary offices who are in charge of the transaction record's guardianship and security. In short, blockchain is a data structure that acts as a distributed database by storing transactions in a logical order and relating them to previous blocks [24–26]. This structure is made up of two parts: first part is a header part and second part is transactions. It also stored the detailed information on the transactions that it holds. As a result, a transaction's source and destination addresses can be linked.

A cryptographic digest generates a unique ID for each block shown in Figure 1.3. The header stores the hash of the block immediately preceding it, allowing us to construct a link between them. As a result, the name blockchain was given to this arrangement.

1.7 Types of Blockchains

Blockchains are of three types i.e. private, public, and consortium, which are shown in Figure 1.4.

1.7.1 Public Blockchain

A public blockchain is an open network that allows people from everywhere to join, transact, and do mining. There are no limitations on any of these variables. As a result,

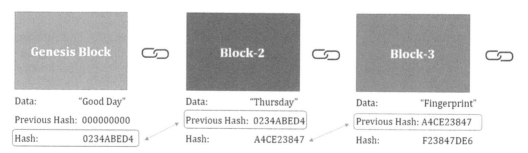

FIGURE 1.3
Blockchain structure.

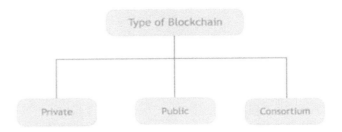

FIGURE 1.4
Types of blockchains.

these blockchains are called permission less blockchains [27–30]. Every member has complete authority to read/write transactions, conduct blockchain audits, and inspect any or every area of the blockchain at any moment.

1.7.2 Private Blockchain

A private blockchain is a sort of blockchain used to give permission of sharing private data among a group of individuals in a single organisation or between many organizations. It is also called as a permissioned blockchain because it can only be accessed by those who have received a special invitation in form of password. To control access, nodes participation is done with a set of rules or by the network administrator.

1.7.3 Consortium Blockchain

A consortium blockchain can be partially private and permissioned blockchain in which block validation are handled by a group of predetermined nodes rather than a single entity. These nodes determine who is allowed to join the network and who is allowed to do mining [31–33]. A multi-signature approach is utilized for block validation, with a block being regarded legitimate only if it is signed by these nodes.

   As a result of the control by a few selected validator nodes, it is a somewhat centralized system, as opposed to the private blockchain, which is totally centralised, and the public blockchain, which is completely decentralized.

1.8 Components of Blockchains

A blockchain has five major components: block, ledger, transaction, consensus mechanism, hash function, and minors, as shown in Figure 1.5.

1.8.1 Block

A block is the fundamental building block of a blockchain. The two main components of a block are the header and transactions.

   The hash of the block is used as nonce and Merkle tree root is used for its further operation. These are the most important fields in the header [34]. To identify the

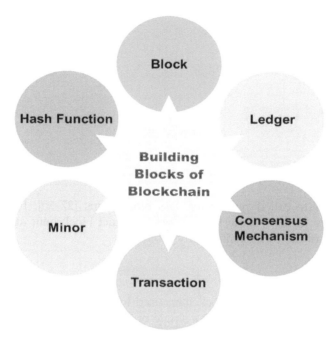

FIGURE 1.5
Blockchain components.

block with its position in the chain, the header hash and block height both are maintained. For a thorough understanding of blockchain, these fields will be shown in Figure 1.6 in detail.

1.8.1.1 Block Header

1.8.1.1.1 Header hash

A header hash is used to recognize the block with a cryptographic digest method and it is also used as an input variable. Each entire node computes a new block when it is received. Then it stores it in a separate database as part of the block metadata.

1.8.1.1.2 Height

The blocks are added to the chain in a chronological order. Using this order, one number is issued to each new block [35]. The height is the difference between the numbers of the last and first blocks. In this case, a fork in the chain happens.

1.8.1.1.3 Hash of the previous block

The previous block's hash is given in the header to allow the block to connect to the one before it.

1.8.1.1.4 Nonce

A nonce is an integer value to adjust the header hash's output. The header hash starts with a series of three 0's, the miner will iterate the nonce until it meets its requirement. When it completed, the process of nodes get a new block. They will only calculate the header hash once to determine if the nonce is valid.

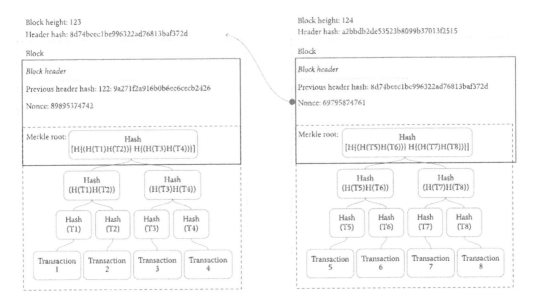

FIGURE 1.6
Structure of block.

1.8.1.1.5 Difficulty

A difficulty is a partial hash collision problem, which means that a hash algorithm always outputs the same assimilate for a given input. If a bit from this input is changed, the resulting hash output will be completely dissimilar. As a result, finding a hash that fulfils this partial collision is dependent on the mining node's processing capability. The collision is created by a mechanism called the nonce.

1.8.1.1.6 Transaction

Transactions are the records of data in the block. A value transfer is referred to as a transaction. It's a collection of inputs and outputs in a nutshell. A node distributes a transaction to all of its neighbors after making it [36]. The nodes that receive the transaction pass it on to their neighbors, and so on, until it reaches every node in the network.

1.8.1.1.7 Merkle trees

A Merkle tree is also often called a binary hash tree. It is one of a complete binary tree with a bit value at each node. An inner node's value is a one-way function of the values of its children.

1.8.2 Ledger

A ledger is used as a database that is shared among multiple sites and institutions among number of people [37,38]. It allows transactions to have public witness. The blockchain is the concept used as distributed ledger by Bitcoin.

1.8.3 Transactions

A value transfer is referred to as a transaction. It's a collection of inputs and outputs in a nutshell. A node distributes a transaction to all of its neighbors after making it. The nodes that receive the transaction pass it on to their neighbors, and so on, until it reaches every node in the network.

On a node-to-node basis, the blockchain allows for the sharing and exchange of information across nodes. This is done through the use of files that carry information from one node to the next, which are generated by a source node and broadcast to the entire network for confirmation. These transactions, which are continuously generated by nodes and subsequently gathered in blocks, represent the current status of blockchain.

After every transaction takes place, the status of the blockchain changes. With so many transactions being made every second, it's difficult to validate and verify the legitimate ones while discarding the fake node [39].

Once a miner collects a transaction result, they save it so that it can be involved in the next block being mined. Once this mined block is added to the chain, the transaction works as public and immutable.

1.8.4 Consensus Mechanism

When nodes start sharing data via a blockchain platform, they lack a centralized unit to regulate and resolve all the problems, as well as to maintain track of the flow of cash and assure a secure exchange to prevent the fraud, such as double spending assaults. To maintain a consistent state, all nodes should agree on a common content update procedure for this ledger, and blocks should not be allowed into the blockchain without majority permission.

This is known as a consensus mechanism because it involves creating new blocks and adding them to the existing ledger for future use. In the case of bitcoin, recipients may justify the outputs numerous times for use in subsequent transactions with signature verification because they appear genuine to individual recipients.

1.8.5 Hash Function

A hash function is defined as a mathematical function that generates a summary of data which is applied to a dataset which produces a one-of-a-kind result. It is one of the most popular advantages to use the value of the hash for verification of data integrity [39]. The hash output size varies depending on the algorithm employed, but what important is that it is always the same size, regardless of the input size.

All network resource transactions are secured using the concepts of keys and digital signatures. The public key cryptography approach is used to generate the keys. A public key that can be shared which is generated, as well as a secret key that only the owner has permission to access that key. The entire transaction requires a signature to be recognised as valid and to demonstrate ownership.

Properties of hash algorithms:

a. One way to identifying the input from hash values must be computationally difficult.

b. Compression: It is just a tiny amount of the data should be represented by the hash size.

c. Calculation ease: It is the hash algorithm should be simple to use.

d. Diffusion: It works when one bit of input is changed, the hash output should be updated by a number of bits close to 50% to avoid reverse engineering of the process.

e. Collision: It should be computationally tough to identify two inputs that produce the same hash.

1.8.6 Minors

Minors are PCs/specialists who attempt to solve a difficult numerical problem in order to study another block. Finding a new block begins with each node transmitting new transactions, following which each node consolidates a group of transactions into a block and begins looking for the block's Proof of Work.

1.9 Privacy and Security Requirements

The security and privacy of any information system are critical requirements. We use concepts like integrity, availability, and confidentiality to describe safety. To establish security, most people utilize a mix of authentication, authorization, and identification.

1.1. **Integrity:** It's the guarantee that the data hasn't been changed with by anyone other than those with the right to do so. Integrity ensures that transactions on the blockchain remain immutable. Integrity is routinely verified using cryptographic processes.

1.2. **Availability:** It ensures that system users have access to it whenever they need it. In other words, when a legitimate user wants the service, it is always available, which needs the communication infrastructure and database. The blockchain achieves this purpose by allowing users to communicate with one another and store blocks in a distributed manner.

1.3. **Confidentiality:** It ensures that unapproved individuals do not gain access to all the information. That is, only those with the necessary rights and obligations will be able to access the data, whether it is being processed or being transported. To ensure this concept, the blockchain makes use of pseudo-anonymization technologies such as hash functions, which hide the identity of users [40].

1.4. **Authorization, Authentication, and Auditing:** This tries to identify the person, who is performing a certain function in a system, as well as determine what privileges that person has and track their usage statistics. The blockchain's structure ensures all three purposes because only users with private keys can make transactions, and all transactions are public and auditable.

1.5. **Non-Repudiation:** It assures that no one can deny the operation of a system. Non-Repudiation proves that a user did something specific, such give money, authorize a transaction, or send a communication. A user can't claim he didn't do something because all transactions are signed. The right to privacy refers to a person's ability to share personal information with others.

1.10 Integration of Blockchain with IoT

Blockchain was first used for monetary and digital currency exchanges, with each member of the BC network executing and storing transactions. Because of the enormous

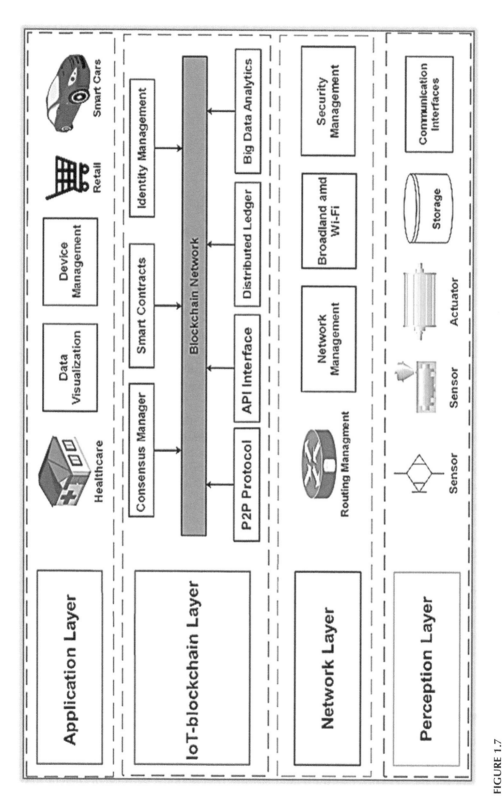

FIGURE 1.7
Layered architecture of IoT and blockchain.

benefits it provides, blockchain is now being used in a variety of settings [41]. One of these areas is the Internet of Things framework. Combining blockchain and IoT can provide a slew of benefits to various IoT applications. The layered architecture of IoT and blockchain is depicted in Figure 1.7.

- In a layered design, all of the gadgets, like as sensors and actuators, collect data from the environment and transfer it to the network layer at the perception layer.
- The network layer is responsible for routing, which allows all connected devices to communicate across the internet.
- The third layer enables all of the blockchain's functions.
- Finally, the uppermost layer, the application layer, offers users with data visualization features and assists them in making suitable decisions based on the obtained data.

The chapter proposes blockchain technology as a solution for IoT data security. Security concerns will continue to grow, necessitating the urgent need for a decentralised and secure innovation to tackle them. Both IoT and blockchain technologies transform concepts and open up new possibilities in their own ways, and it is feasible to create applications that combine the benefits of both [42]. The BC invention guarantees data security by preventing intruders from controlling or feeding bogus data without breaking the chain.

1.11 Conclusion

This chapter presents a recent literature study and analysis on blockchain in the Internet of Things. We outline five critical components, as well as in designing and problems, that should be taken into account while developing blockchain architecture for IoT. This chapter also identifies the deficiency that block the development of a secure blockchain framework for IoT. This chapter also proposed uses for blockchain innovation in a variety of social insurance settings, including critical care, restorative data inquiry, and associated wellness. In this chapter, also discussed is how retaining a permanent and simple document, which displays all of the events that occurred on the device, can help to improve and stimulate the management of therapeutic records. As a feature of future work, we may need to develop the structure of a patient record and its metadata, leveraging the semantics of social insurance data, including the possibility of sharing radiology pictures, which is much more difficult. We plan to test our framework with the facts of real patients because we are working in collaboration with a health facility. Our long-term goal is to look into the unique conditions proposed in the research, such as related well-being.[1,2,3]

Notes

1 at 1 bar.
2 at 15 °C.
3 at 200 bar.

References

[1] Yousuf, A. M., E. M. Rochester, B. Ousat and M. Ghaderi, "Throughput, coverage and scalability of LoRa LPWAN for Internet of Things," IEEE/ACM 26th International Symposium on Quality of Service (IWQoS), 2018, pp. 1–10.

[2] Lee, C. H. and K.-H. Kim, "Implementation of IoT System Using Blockchain with Authentication and Data Protection," International Conference on Information Networking (ICOIN), 2018, 10.1109/icoin.2018.8343261

[3] Babar, S., A. Stango, N. Prasad, J. Sen and R. Prasad, "Proposed Embedded Security Framework for Internet of Things (iot)," In 2nd International Conference on Wireless Communication, Vehicular Technology, Information Theory and Aerospace & Electronics Systems Technology (Wireless VITAE), 2011, pp. 1–5, IEEE.

[4] Das, D. and S. Sarkar, "Machine-to-machine Learning based framework for ad-hoc IOT ecosystems," International Conference on Computational Techniques, Electronics and Mechanical Systems (CTEMS), 2018, 10.1109/ctems.2018.8769148

[5] Lei, K., M. Du, J. Huang and T. Jin, "Groupchain: Towards a Scalable Public Blockchain in Fog Computing of IoT Services Computing." *IEEE Transactions on Services Computing* 13 (2018) (2): 252–262. 10.1109/tsc.2019.2949801

[6] Li, S., T. Qin and G. Min, "Blockchain-Based Digital Forensics Investigation Framework in the Internet of Things and Social Systems," *IEEE Transactions on Computational Social Systems* 6(2019) (6): 1–9. 10.1109/tcss.2019.2927431

[7] Madumidha, S., P. S. Ranjani, U. Vandhana and B. Venmuhilan, "A theoretical implementation: Agriculture-food supply chain management using blockchain technology," 2019 TEQIP III Sponsored International Conference on Microwave Integrated Circuits, Photonics and Wireless Networks (IMICPW), 2019. 10.1109/imicpw.2019.8933270

[8] Karampatzakis, D., G. Avramidis, P. Kiratsa, I. Tseklidis and C., Oikonomidis, "A Smart cargo bike for the physical internet enabled by RFID and LoRaWAN," Panhellenic Conference on Electronics & Telecommunications (PACET), 2019, 10.1109/pacet48583. 2019.8956282

[9] Zhu, K., Z. Chen, W. Yan and L. Zhang, "Security Attacks in Named Data Networking of Things and a Blockchain Solution." *IEEE Internet of Things Journal* 6 (2018) (6). 10.1109/jiot. 2018.2877647

[10] Mick, T., R. Tourani and S. Misra, "LASeR: Lightweight Authentication and Secured Routing for NDN IoT in Smart Cities." *IEEE Internet of Things Journal*, 5 (2018) (2): 755–764. 10.1109/jiot.2017.2725238

[11] Zhang, Y. and J. Wen, "An IoT electric business model based on the protocol of bitcoin," 18th International Conference on Intelligence in Next Generation Networks, 2015. 10.1109/ icin.2015.7073830

[12] Abeyratne, S. A. and R. P. Monfared, "Blockchain Ready Manufacturing Supply Chain Using Distributed Ledger." *International Journal of Research in Engineering and Technology* 5 (2016) (9): 1–10 .

[13] Bao, Z., W., Shi, D., He, and K. K. R., Chood. IoTChain: A three-tier blockchain-based IoT security architecture. arXiv:1806.02008 (2018).

[14] "Comodo News and Internet Security Information. Bitcoin Phishing Attack | Hacking Methods Used for Cryptowallets," https://blog.comodo.com/comodo-news/bitcoin-phishing-attack-on-cryptowallet-owner/, (2018).

[15] Conoscenti, M., A. Vetro and J. C. De Martin, "Blockchain for the Internet of Things: A Systematic Literature Review," In: Proceedings of IEEE/ACS International Conference on Computer Systems and Applications, AICCSA, 2017.

[16] Hang, L and D. H. Kim, "Design and Implementation of an Integrated IoT Blockchain Platform for Sensing Data Integrity." *Sensors* 19 (2019) (10): 2228. 10.3390/s19102228

[17] Fernandez-Carames, T. M. and P. Fraga-Lamas, "A Review on the Use of Blockchain for the Internet of Things." *IEEE Access* 6 (2018): 32979–33001. 10.1109/access.2018.2842685

[18] Sharma, P. K., S. Singh, Y.-S. Jeong and J. H. Park, "DistBlockNet: A Distributed Blockchains-Based Secure SDN Architecture for IoT Networks." *IEEE Communications Magazine* 55 (2017) (9): 78–85. 10.1109/mcom.2017.1700041

[19] Tselios, C., I. Politis and S. Kotsopoulos, "Enhancing SDN Security for IoT-Related Deployments Through Blockchain," 2017 IEEE Conference on Network Function Virtualization and Software Defined Networks (NFV-SDN), 2017. 10.1109/nfv-sdn.2017.8169860

[20] Kravitz, D. W. and J., Cooper, "Securing User Identity and Transactions Symbiotically: IoT Meets Blockchain." *Global Internet of Things Summit (GIoTS)*, (2017): 1–6. 10.1109/giots. 2017.8016280

[21] Samaniego, M. and R., Deters, "Blockchain as a Service for IoT. IEEE International Conference on Internet of Things (iThings) and IEEE Green Computing and Communications (GreenCom) and IEEE Cyber," *Physical and Social Computing (CPSCom) and IEEE Smart Data (SmartData)*, (2016). 10.1109/ithings-greencom-cpscom-smartdata.2016.102

[22] Biswas, K. and V. Muthukkumarasamy, "Securing Smart Cities Using Blockchain Technology," 2016 IEEE 18th International Conference on High Performance Computing and Communications; IEEE 14th International Conference on Smart City; IEEE 2nd International Conference on Data Science and Systems (HPCC/SmartCity/DSS), 2016. 10.1109/hpcc-smartcity-dss.2016.0198

[23] Yuan, Y. and F.-Y. Wang, "Towards blockchain-based intelligent transportation systems," 2016 IEEE 19th International Conference on Intelligent Transportation Systems (ITSC), 2016, 10.1109/itsc.2016.7795984

[24] Varshney, G. and H. Gupta, "A Security Framework for IOT Devices Against Wireless Threats," 2017 2nd International Conference on Telecommunication and Networks (TEL-NET), 2017. 10.1109/tel-net.2017.8343548

[25] Kouzinopoulos, C. S., K. M. Giannoutakis, K. Votis, D. Tzovaras, A. Collen, N. A. Nijdam, … S. Katsikas, "Implementing a Forms of Consent Smart Contract on an IoT-Based Blockchain to Promote User Trust." *Innovations in Intelligent Systems and Applications (INISTA)*, (2018): 1–6. 10.1109/inista.2018.8466268

[26] Yue, L., H. Junqin, Q. Shengzhi and W. Ruijin, "Big Data Model of Security Sharing Based on Blockchain," 3rd International Conference on Big Data Computing and Communications (BIGCOM), 2017. 10.1109/bigcom.2017.31

[27] Abdallah, E. G., H. S. Hassanein, and M. Zulkernine, "A Survey of Security Attacks in Information-Centric Networking." *IEEE Communications Surveys and Tutorials* (2015): 1441–1454.

[28] Afanasyev, A., P. Mahadevan, I. Moiseenko, E. Uzun and L. Zhang, "Interest Flooding Attack and Countermeasures in Named Data Networking," In IFIP Networking Conference, 2013, pp. 1–9.

[29] Ogiela, M. R. and M. Majcher, "Security of Distributed Ledger Solutions Based on Blockchain Technologies," 2018 IEEE 32nd International Conference on Advanced Information Networking and Applications (AINA), 2018. 10.1109/aina.2018.00156

[30] Liu, C., Y. Zhang and H. Zhang, "A Novel Approach to IoT Security Based on Immunology," 2013 Ninth International Conference on Computational Intelligence and Security, 2013. 10.1109/cis.2013.168

[31] Chze, P. L. R. and K. S. Leong, "A secure multi-hop routing for IoT communication," 2014 IEEE World Forum on Internet of Things (WF-IoT), (2014). 10.1109/wf-iot.2014.6803204

[32] Perez, S., J. A. Martinez, A. F. Skarmeta, M. Mateus, B. Almeida and P. Malo, "ARMOUR: Large-scale Experiments for IoT Security and Trust," *2016 IEEE 3rd World Forum on Internet of Things (WF-IoT)*, 2016. 10.1109/wf-iot.2016.7845504

[33] Han, J.-H., Y. Jeon and J. Kim, "Security Considerations for Secure and Trustworthy Smart Home System in the IoT Environment," 2015 International Conference on Information and Communication Technology Convergence (ICTC), 2015. 10.1109/ictc.2015.7354752

[34] Giaffreda, R., Capra, L. and Antonelli, F., "A Pragmatic Approach to Solving IoT Interoperability and Security Problems in an eHealth Context," *IEEE 3rd World Forum on Internet of Things (WF-IoT)*, 2016. 10.1109/wf-iot.2016.7845452

[35] Ray, S., S. Bhunia, Y. Jin and M. Tehranipoor, "Security Validation in IoT Space," IEEE 34th VLSI Test Symposium (VTS), 2016. 10.1109/vts.2016.7477288

[36] Tellez, M., S. El-Tawab and M. H. Heydari, "IoT Security Attacks Using Reverse Engineering Methods on WSN Applications," *IEEE 3rd World Forum on Internet of Things (WF-IoT)*, 2016. 10.1109/wf-iot.2016.7845429

[37] Johnston, S. J., M. Scott and S. J. Cox, "Recommendations for securing Internet of Things Devices Using Commodity Hardware," *IEEE 3rd World Forum on Internet of Things (WF-IoT)*, 2016. 10.1109/wf-iot.2016.7845410

[38] Baldini, G., A. Skarmeta, E. Fourneret, R. Neisse, B. Legeard and F. Le Gall, "Security Certification and Labelling in Internet of Things," *IEEE 3rd World Forum on Internet of Things (WF-IoT)*, 2016. 10.1109/wf-iot.2016.7845514

[39] Son, M. and H. Kim, "Blockchain-Based Secure Firmware Management System in IoT Environment," 21st International Conference on Advanced Communication Technology (ICACT), 2019. 10.23919/icact.2019.8701959

[40] Ali, A. and N. Singh, "QoS Analysis in MANETs Using Queueing Theoretic Approaches A review." *International Journal of Latest Trend in Engineering and Technology (IJLTET)* 7 (2016) (1): 120–124.

[41] Parveen, N., A. Ali and A. Ali, "IOT based automatic vehicle accident alert system," 2020 IEEE 5th International Conference on Computing Communication and Automation (ICCCA), 2020, pp. 330–333, Greater Noida, 10.1109/ICCCA49541.2020.9250904. (Scopus Indexed).

[42] Sachdeva, S. and A. Ali, "A Hybrid Approach Using Digital Forensics for Attack Detection in a Cloud Network Environment." *International Journal of Future Generation Communication and Networking* 14 (2021) (1): 1536–1546.

2

The Study of Consensus Algorithms in Blockchain

Debatosh Pal Majumder

*Department of CSE, Netaji Subhash Engineering
College, Kolkata, India*

CONTENTS

2.1 Introduction

Blockchain [1,2] is a distributed chain of nodes. Blockchain acts as an immutable [3] ledger that can track transactions [4,5] and assets. Blockchains can hold both tangible and intangible items and can record virtually anything of value. Digital transactions can also be performed on a blockchain network. Performing digital transactions on a blockchain network reduces risk and cost. Bitcoin [6], one of the most widely uses digital currencies [7], uses blockchain as its fundamental framework. Since all the transactions are encrypted [8] and are immutable, cryptocurrencies are safer than traditional methods. Blockchain also acts as a public ledger [9], which makes it open and accessible to anyone using the currency.

In 1982, cryptographer David Chaum proposed a technology similar to blockchain. Stuart Haber and W. Scott Stornetta published further work on cryptographically secure blockchains in 1991. They wanted to implement a mechanism to prevent document timestamps from being forged. In 1992, Haber, Stornetta, and Dave Bayer improved the architecture by using a Merkle tree that can combine many document certificates into one block.

DOI: 10.1201/9781003203957-3

Since 1995, hashes of document certificates have been published weekly in the *New York Times* under the name Surety.

Satoshi Nakamoto developed the first blockchain in 2008. Nakamoto significantly enhanced the architecture by incorporating a difficulty parameter to steady the pace at which blocks are added to the chain and employing a Hashcash-like approach to timestamp blocks without having them be signed by a trusted party. The next year, Nakamoto implemented the idea as a core component of the cryptocurrency Bitcoin, where it serves as the public record for all network transactions.

A blockchain network is a chain of nodes. Each node contains the data, the hash of the previous node, its hash, and nonce. A "number only used once," also known as a nonce, is a number that is used to meet the difficulty criteria. A blockchain can hold any type of data like digital currencies, ownership proof, etc. The hash in each block is generated by encryption algorithms like sha256, etc. and each hash is unique. A small change in the block will alter the hash completely. This hash acts as a method to stop a malicious group from altering the data. The block also contains information about the parties involved in the transaction. Since a slight change in the data changes the hash completely, anybody trying to modify the chain the entire network becomes invalid. Each node also contains the hash of the previous node, which makes altering the chain harder. Thus, hashing restricts the attempts to alter the chain. Section 1.1 gives the importance of blockchain technology.

2.1.1 Importance of Blockchain Technology

In the blockchain network, the new node contains the hash of the previous node, which makes it even more secure and the node is encrypted, making the blockchain much more reliable and secure than traditional ledger maintaining systems. As the name suggests, a blockchain is created by a network of computers that are linked together to authenticate blocks, which are added to the ledger to form a chain. Each block in the blockchain network contains complex hashes and information which cannot be altered once created. Blockchain is immune to tampering and hacking because it is immutable and incorruptible. Blockchain also improves traceability, as every time exchange of items is recorded on the blockchain, an audit trail is provided to trace the origin of goods. And due to its decentralized nature, blockchain does not require intermediaries for any transactions, including payments and real estate. The unique feature of blockchain is that it is a public ledger visible to all, which adds an unprecedented level of accountability to the financial system and businesses, ensuring that each sector is responsible for the development of the community. Section 2 gives mostly used consensus [10] mechanisms, Section 3 gives a relative study, and Section 4 concludes.

2.2 Consensus Mechanisms

As previously stated, a hash is a useful tool for detecting attempts to modify data in blocks. However, the hash algorithm alone is insufficient to assure the security [11] of a blockchain. The network employs consensus [12] procedures to thwart efforts to change the blockchain and assure security.

A consensus [13] mechanism is an algorithm that approves transactions or records on a decentralized ledger while rejecting false or fraudulent records. When additional blocks are added to the current chain of blocks, the algorithm is executed. The notion is that by requiring a certain amount of work, malevolent actors will abstain from tampering with the ledger since the effort will be unprofitable. The very first application of proof of work [14] was to screen email spam.

Adam proposed Hashcash, a proof-of-work technology. Back in 1997, it required email senders to create and attach stamps to email headers to demonstrate to recipients that they used CPU power to compose emails. These stamps are one-way encryption schemes that are simple for the recipient to verify but difficult for the sender to manufacture. Spammers would be hesitant to send out big amounts of emails under this paradigm because it becomes unprofitable to require a large volume of CPU resources to manufacture stamps. Regular users, however, may still send a single email at a reasonable fee.

Consensus techniques in the blockchain realm are commonly referred to as mining and staking activities; they are typically seen as a means to issue new currency. Their primary goal, however, is to safeguard the decentralized [15] network, while payments in the form of coins provide an additional economic incentive for employees to keep the network running. The subsections explained proof of work, proof of stake, delegated proof of stake, proof of capacity, proof of elapsed time, proof of authority, and proof of activity in Section 2.1 to Section 2.7, respectively.

2.2.1 Proof of Work

The proof-of-work method was developed in 1993. Malicious activities like denial of service attacks and other network spams were prevented by this method. The algorithm required some work from the user in terms of processing power to validate. In 2009, Bitcoin used the proof-of-work algorithm to validate transactions in its network. To add new blocks, users have to use their computational power to solve a cryptographic puzzle [16] and complete the proof of work validation. This innovative use of the proof-of-work algorithm has been adopted by many cryptocurrencies and it has become the widely used consensus algorithm.

Users on a network exchange digital tokens. All transactions are collected into blocks by a decentralized ledger. However, caution should be exercised to confirm transactions and organize blocks. This obligation falls on specific nodes known as miners, and the process is known as mining. The way that Bitcoin chooses to implement proof of work, is in a quite simple, maybe deceptively simple hashing puzzle. The Bitcoin protocol requires the nodes to attach a random number called a nonce to the transaction data in the block while generating their blocks. Simply put, the hashing puzzle is as follows: Take a hash of the block plus the nonce, and find at least X leading zeros in the resultant hash. It must be obtained by trial and error. Repeatedly hash the ever-changing block with different nonces until you have the required amount of zeros. When the miner ultimately discovers the correct answer, the node broadcasts it to the whole network at the same moment, in exchange for a Bitcoin award offered by the proof-of-work protocol. Changing a block necessitates regenerating all successors and redoing the work they contain, which amounts to computing the complete chain of difficult mathematical problems, which is nearly impossible. This protects the blockchain against manipulation. Figure 2.1 gives the block diagram of proof of work.

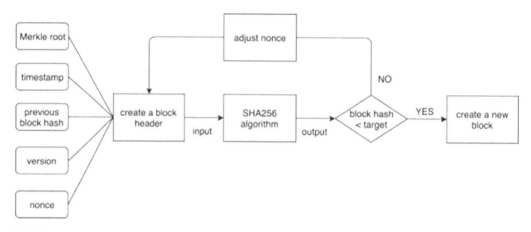

FIGURE 2.1
Block diagram of proof of work pros.

Pros

1. Proof of work is a time-tested consensus mechanism that has maintained Bitcoin and Ethereum [17] safe and decentralized for years.
2. It is quite simple to implement in comparison to proof of stake.
3. Unlike proof of stake, proof of work does not require you to have coins for mining. Anyone can start mining without having to stake coins.

Cons

1. As high computational power is required for solving the hashing problem, specialized hardware is required to increase the initial cost.
2. Mining as a resource-intensive process consumes a lot of energy, which is bad for the environment.
3. Because of the increased computing required, mining pools may come to dominate the mining game, resulting in centralization and security problems.

2.2.2 Proof of Stake

The second major consensus protocol is called proof of stake. In proof of stake, the right to mine the blocks is given out not randomly, but proportionally. So to mine a new block, a person has to stake some coins in the network, and the probability of mining the block is proportional to the number of coins staked to the network. Hence, the mining power of the miner depends on the number of coins he/she owns. The proof-of-stake consensus mechanism was developed to solve the scalability [18] and environmental sustainability problems raised by the proof-of-work protocol.

As implied by the name, nodes on a network stake a certain amount of cryptocurrency to be considered candidates to validate the new block and collect the fee associated with it. Then, an algorithm selects the node that will validate the new block from a pool of candidates. To make the selection fair to everyone on the network, this algorithm combines the amount of stake with other factors such as coin-age-based selection and the random block selection process. In coin-age-based selection, the method keeps track of

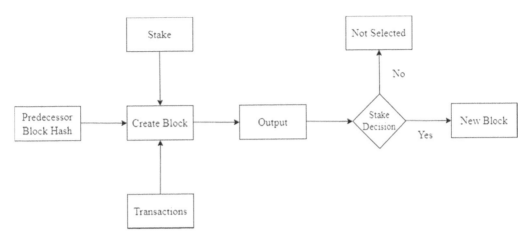

FIGURE 2.2
Block diagram of proof of stake pros.

how long each validator candidate node remains a validator. The older the node, the more likely it is to become the new validator. In the random block selection process, the validator is determined by combining the lowest hash value with the largest stake and the node with the best weighted combination of these is designated as the new validator. Figure 2.2 gives a block diagram of proof of stake.

Pros

1. Energy is conserved since nodes are not competing with one another to add a new block to the blockchain.
2. More decentralized compared to the proof-of-work mechanism
3. An individual attempting to attack the network will need to hold 51% of the stakes.

Cons

1. PoS is still in its early stages.
2. If a group of validator candidates pool their resources and hold a considerable portion of the total cryptocurrency, they will have a better chance of becoming validators. Increased chances lead to more choices, which leads to more forging reward earning, which leads to holding a large currency stake. As a result, the network may become centralized over time.

2.2.3 Delegated Proof of Stake

This consensus mechanism is an advanced and democratic version of the more widely used proof-of-stake mechanism. Delegated proof of work was developed in 2014 by Daniel Larimer and many cryptocurrencies like Bithshares, Steem, Ark, Lisk, etc. use this consensus mechanism.

Participants still stake coins in a delegated proof-of-stake system like in proof of stake. Rather than being accountable for validation themselves, stakeholders delegate that task to a delegate, with groups of stakeholders, who are then responsible for establishing consensus among themselves. Delegated proof-of-stake delegates are elected based on their reputation and perceived trustworthiness, and it is believed that the system incentivizes good conduct among delegates because the community has the power to vote them out and replace them at any moment. This protocol allows stakeholders to select a few participants who can act as delegates, validating new blocks to be added, overseeing the entire consensus-building process, and being rewarded for doing so. The probability of being selected is proportional to the number of coins the user owns. Mining rewards are distributed among all participants responsible for adding new blocks to the blockchain network.

Pros

1. Delegated proof-of-stake networks has strong protection from double-spend attack.
2. As the entry threshold is so low, delegated proof of stake is often regarded as the most decentralized way to the consensus process.
3. Delegated proof of stake is much more scalable than the proof of work method as they don't require high computational power.
4. Delegated proof-of-stake blockchains are faster than proof of work and proof of stake consensus mechanisms.
5. As a delegated proof of stake doesn't require high computational power, it consumes less power and is environment friendly.

Cons

1. A small number of witnesses might lead to network centralization.
2. The DPoS blockchain is vulnerable to the shortcomings of traditional real-world voting.
3. Delegators must be well aware and appoint honest witnesses for the network to operate and make decisions effectively.

2.2.4 Proof of Capacity

Proof-of-capacity mining is a very new mining method that is presently being employed by cryptocurrencies like SpaceMint, Permacoin, and Burstcoin. Many people believe that proof of capacity is a viable alternative to the mining techniques currently in use. It enables mining by utilizing hard disc space. A strategy like this has a lot of advantages, and it's something that many projects are considering right now. Miners employ computer storage rather than the more prevalent and energy crunching proof-of-work approach, which includes continuous computation. This type of mining is similar to the proof of work but in a condensed way where you compute once and store the results on hard drive space. Then mining simply involves reading through your cache and the storage is idle most of the time and only goes through the plot files for a few seconds every block.

To achieve this method of mining, miners create plots that contain all of the computations required to make the blocks, this process is known as plotting. Plotting is the only aspect of the proof of capacity mining process that uses energy, and it uses substantially less than proof of work. Miners use available space in a plot file to store nonces generated

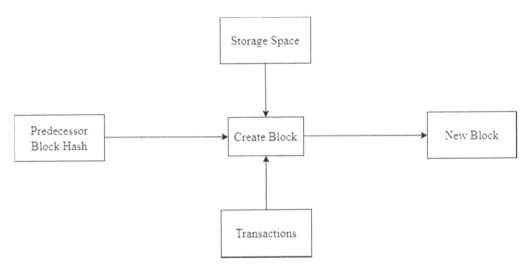

FIGURE 2.3
Block diagram of proof of capacity.

by repeated hashing of data. Setting up this plot is a one-time expense, and it may then be used to show validators that the space has been assigned, allowing you to mine. This consensus process is more energy-efficient since it works with any hard drive, even mobile devices, and users may reuse the hard drive once they finish mining. Users do not need to change their gear regularly unless they wish to increase the amount of hard disk space available for mining. Figure 2.3 illustrates the proof of capacity.

Plotting uses a Shabal cryptographic algorithm, which is used to create the plots.

Pros

1. Any hard drive, even mobile devices, can be used in this method in contrast to high computation equipment used in the proof-of-work mechanism.

2. Mining with hard drives consumes 30 times less electricity than using ASIC miners used in the proof-of-work mechanism.

3. As everyone has a hard disk, this method is more decentralized than methods like proof of stake.

4. As any type of hard drive can be used in this mechanism, miners won't need to upgrade their equipment regularly.

5. Once mining is finished, the hard disk may be wiped and returned to its intended use.

Cons

1. This method of consensus has been adopted by a very small community.

2. Hackers will try to exploit the situation. Many computers throughout the world are presently infected with mining malware. In proof of work, mechanism malware can be easily detected by identifying the cause of the decreased performance of the system. Proof of capacity, on the other hand, makes determining if your extra hardware space is being used for illicit purposes much more difficult.

3. Even though proof-of-capacity mining offers lower entry barriers, users may opt to buy larger hard drives and mine the majority of the money.

2.2.5 Proof of Elapsed Time

A new consensus mechanism is designed to improve proof-of-work consensus and provide a new alternative to permission blockchain networks. Proof of elapsed time consensus is used in private or permissive blockchain networks such as the Hyperledger Sawtooth modular structure. Each node is assigned a mining waiting period by the network, and the node with the shortest waiting period wins first. Proof of elapsed time consensus was developed by Intel for use in permission blockchains. Proof of elapsed time requires special Intel chips. The method requires a special set of CPU instructions known as Intel software guard extensions. Hosts interested in joining the network should download the trusted code and run SGX in the trusted runtime environment. Proof of elapsed time consensus has now become a popular consensus mechanism for permission blockchains.

The miners must first join the network and get a certificate of membership. The nodes must wait a set period after joining the network, which is determined at random. Before mining a new block into the blockchain, the miner must wait at least the length of time specified. In this system, the miner with the shortest period is chosen to mine the blocks for that round. The mechanism is generally fair, selecting miners with a high degree of randomness. It does not consume a lot of electricity, as miners may sleep while waiting for their turn. The method uses SGX protocol, allowing the logical isolation of CPU memory that cannot be accessed or altered. These components, sometimes known as enclaves, can perform isolated instructions and memory encryption. Only enclaves have access to edit the contents of this compartment. The code is encrypted and cannot be viewed outside of the enclave, making it extremely safe for the processes that take place within it.

Pros

1. Rather than being resource-intensive like the proof-of-work mechanism, it allows a miner's CPU to sleep and transition to other jobs for a set period, enhancing efficiency.
2. Many additional network requirements are met by the method of running trustworthy programs in a safe environment. It guarantees that the trusted code executes within the secure environment and that it cannot be changed by anybody else.

Con

1. As SGX is wholly manufactured by Intel, the consensus model's dependence extends to Intel as a firm, a third party. Such dependency contradicts the new paradigm that cryptocurrencies are seeking to accomplish through blockchain networks.

2.2.6 Proof of Authority

This new consensus mechanism, primarily for private [19] networks, provides a reputation-based approach that eliminates the computationally intensive need to add new blocks to the blockchain network. The proof of authority was developed by Ethereum co-founder Gavin

Wood in 2017. This consensus mechanism has recently received a lot of attention for the scalability of blockchain networks. Proof of authority is a modified variant of proof of stake in which, instead of a monetary stake, a validator's identification serves as the stake. Because the proof of authority consensus mechanism utilizes the value of identities, block validators are not staking coins but rather their reputation. As a result, proof-of-authority blockchains are protected by validating nodes that are randomly chosen as trustworthy organizations. The proof-of-authority concept is based on a small number of block validators, which makes it a highly scalable system. Blocks and transactions are checked by pre-approved individuals who serve as system moderators. Proof of authority is a consensus method aimed at businesses or private groups who wish to create chains that are effectively closed and do not require involvement from the broader public. This mechanism is much more efficient since it eliminates resource-intensive mining processes, but validators need to maintain node integrity. As reputation and identity are at stake, the system encourages users to act honestly through the proper functioning of the network.

Amid the push to abandon proof of work, proof-of-stake algorithms have emerged as one of the most popular consensus solutions. While the benefits of proof of stake are obvious, there is one important disadvantage that is sometimes neglected. Proof-of-stake algorithms are based on the concept that those who have staked tokens in a network will be encouraged to behave in the network's best interests or risk losing their investment. As a result, it is natural to believe that the more a person's investment in the network, the more driven they are to ensure its success. This assumption, however, fails to account for the fact that, while identical stakes may be equal in monetary worth, they may not be equally appreciated by their holders. Proof of authority seeks to help with this. Instead of tokens, network participants stake their identities, which is the idea of the algorithm. This implies that, unlike many other blockchain protocols, validators in PoA systems are known entities that risk their reputations in exchange for the ability to validate blocks. (Figure 2.4).

Pros

1. Risk tolerance is high as long as 51% of the nodes are not malicious.

2. High transaction rate.

3. Consumes less computational power than consensus methods like proof of work.

4. The duration between new blocks being created is predictable.

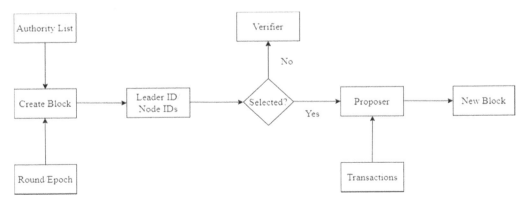

FIGURE 2.4
Block diagram of proof of authority.

Cons

1. Proof of authority is not a decentralized system; rather, it is an attempt to improve the efficiency of centralized institutions.
2. Anyone can see the proof-of-authority validators.

2.2.7 Proof of Activity

Proof of activity represents a blockchain consensus algorithm that is used to make sure that all transactions registered on a blockchain are genuine and that all miners reach a consensus. It also serves as a protection against attacks on the underlying blockchain. Proof of activity is a combination of two other blockchain consensus algorithms, proof of work and proof of stake. Proof of activity is used in blockchain networks like Dcred, Espers, and CoinBureau.

The first phase employs the proof-of-work technique, in which miners compete with their processing power to become the first one to solve a complex task to generate a new block for the blockchain. Once the block is mined, the system proceeds to the second step, in which participants are chosen at random from the network, similar to the proof of stake. Just as in normal proof of stake, the odds of getting chosen rise with the number of coins a member holds in that network. The personnel chosen are then tasked with confirming and signing the created block. Depending on the implementation, they may just need to validate or sign the block to certify their authenticity. When the group of validators signs the new block, the status of a complete block will be achieved. With the new block being identified, it will be added to the existing blockchain network, with the transactions of the new block being recorded. The first miners and the various validators who played a role in contributing to the new block will be the ones who receive the reward. Proof of activity offers a second layer of security against 51% of attacks since an attacker would potentially require 51% or more of the network's total mining power as well as 51% or more of the coins staked in the network to effectively pull off the attack.

Pro

1. Only if a specified number of verifiers acknowledge the block as legitimate can it be permanently attached to the blockchain. In other words, introducing a modified block into the system is difficult.

Con

1. As proof of activity employs proof of work in the first phase, it consumes a lot of computational power.

2.3 Relative Study

Scalability is one of the most critical issues in blockchain, and it has been a focus of both business practitioners and academic researchers since Bitcoin's inception. Because of its benefits, blockchain systems are well suited for circumstances in which mutually distrusting parties must trade value digitally. These benefits are not free; a blockchain is a sluggish, immutable database with extremely high redundancy, which means it is costly to maintain and difficult to grow. However, a scalable blockchain is sometimes misunderstood as just another phrase for a blockchain with a high transaction per second. However, the term *scalability* has a far larger variety of connotations in the context of blockchain. For example, every increase to Bitcoin in terms of throughput, latency, start-up time, or cost per transaction is referred to as scaling, and the resulting blockchain system is referred to as scalable. In the blockchain, the phrase scalable is a comparison concept. When a blockchain system is referred to as scalable, it means that it achieves a greater TPS than previous systems by improving its consensus method and/or tweaking some system parameters.

There are some potential solutions for blockchain scalability. One scaling concept is to incorporate more transactions in a block by increasing the block size or using fewer bytes to represent information in the block. To prevent network spamming, networks like Ethereum and Bitcoin have an upper limit on block size. This limit can be raised to accommodate more transactions. We can obtain higher throughput by reducing the amount of information in each block by utilizing more efficient hashing algorithms and data structures. However, increasing block size means increasing block propagation time, which means the blocks will take longer to verify.

We can accelerate transactions by reducing the time it takes to add a new block to the network. This time is dictated by the block difficulty level in proof-of-work systems. Shortening the block generation time, on the other hand, will increase the number of mined blocks that do not make it to the main chain due to competition, resulting in further computation waste. This increases network bandwidth while reducing network security by lowering the effective hash rate.

We can use various consensus protocols to increase the network's scalability. Proof of stake utilizes less power than proof of work. Proof of stake will also increase the frequency of block additions since miners will no longer have to spend time, solving cryptographic puzzles. While proof of stake is more scalable than proof of work, it has several disadvantages.

To improve scalability, we can employ the off-chaining strategy. The off-chaining strategy, as the name implies, incorporates the off-chaining portion of the processing. The burden on the blockchain network and transaction processing costs would be reduced as a result. Only a few nodes are permitted to calculate the difficult proof in off-chaining. Before computing the proof, the participating nodes must provide a deposit. A group of nodes verifies the proof after it has been collected. If the proof is valid, the deposit and incentive are returned to the submitter. If not, the deposit is fortified.

Sharding [20] may be used to boost scalability in a blockchain network. Because the blockchain state is partitioned into shards, a node does not need to rely on the complete data set to validate and perform a transaction. Transactions will be routed to nodes depending on the shards they affect, and all nodes will work in parallel, lowering total time.

As a layer atop the blockchain architecture, subchain architecture may be employed. Subchain architecture features a tree-like hierarchical structure with a root chain and several child chains. It operates based on strong and weak blocks. Weak blocks are those that meet a target of lower difficulty. As soon as a weak block is discovered, it is distributed to other miners. Miners pile up weak blocks until a block with the specified difficulty goal is discovered (strong block). When the goal is reached, the child chain is closed and reconnected to the root chain, decreasing the average first confirmation time of transactions. Table 2.1 gives a relative study of consensus mechanisms.

TABLE 2.1

A Relative Studies of Consensus Mechanisms

Property	Adversary Tolerance	Power Consumption	Scalability	Application Type	Transaction Rate (TPS)	Examples
Proof of Work	<25% computing power	High	High	Permissionless	7 (Bitcoin) 25 (Ether)	Bitcoin, Ether
Proof of Stake	<50% stake	Medium	High	Both	50000 (Solana)	Solana, Cardano, Polkadot
Delegated Proof of Stake	<51% validators	Low	Medium	Both	100000 (BitShares)	Bitshares, Steem, Ark, Lisk
Proof of Capacity	<25% computing power	Medium	Low	Permissionless	80 (Burstcoin)	SpaceMint, Permacoin, Burstcoin
Proof of Elapsed Time	<25% computing power	Low	High	Both	>1000 (Hyperledger Sawtooth)	Hyperledger Sawtooth
Proof of Authority	n/a	Low	High	Permissioned	n/a	Microsoft Azure
Proof of Activity	<25% computing power	High	High	Permissioned	14 (Decred)	Dcred, Espers, CoinBureau

2.4 Conclusion

Blockchain technology is projected to completely transform the global financial and banking sectors. Many banks have already begun developing blockchain applications or are searching for methods to do so. A comparison of consensus procedures is provided in this chapter. The consensus process guarantees that the blockchain system runs smoothly. A consensus mechanism is used to achieve agreement among nodes on a certain value or transaction.

Based on detailed research of protocol design and the impact of blockchain networks, we gave a perspective on the expanding usage of blockchain networks in numerous fields. This chapter is meant to be a useful guide for pursuing intriguing research avenues that may lead to a better knowledge of the blockchain consensus process and fascinating outcomes in adjacent domains.

References

[1] Natoli, C., J. Yu, V. Gramoli and P. Veríssimo, "Deconstructing Blockchains: A Comprehensive Survey on Consensus, Membership and Structure," (2019). https://arxiv.org/abs/1908.08316

[2] Castro, M. and B. Liskov, "Practical Byzantine Fault Tolerance and Proactive Recovery," *ACM Transactions on Computer Systems* 20 (2002) (4): 398–461.

[3] Sankar, L. S., M. Sindhu and M. Sethumadhavan, "Survey of Consensus Protocols on Blockchain Applications," 2017 4th International Conference on Advanced Computing and Communication Systems (ICACCS), 2017, pp. 1–5.

[4] Nguyen, G.-T. and K. Kim, "A Survey About Consensus Algorithms Used in Blockchain." *Journal of Information Processing Systems* 14 (2018): 101–128.

[5] Xiao, Y., N. Zhang, W. Lou and Y. T. Hou, "A Survey of Distributed Consensus Protocols for Blockchain Networks." *IEEE Communications Surveys and Tutorials* 22 (2020): 1432–1465.

[6] Nakamoto, S., "Bitcoin: A Peer-to-Peer Electronic Cash System," Self-published Paper May 2008. [Online]. Available: https://Bitcoin.org/Bitcoin.pdf

[7] Tschorsch, F. and B. Scheuermann, "Bitcoin and Beyond: A Technical Survey on Decentralized Digital Currencies," *IEEE Communications Surveys Tutorials* 18 (2016) (3): 2084–2123.

[8] Bosamia, M. and Patel, D. "Comparisons of Blockchain based Consensus Algorithms for Security Aspects." *International Journal on Emerging Technologies* 11 (2020): 427–434.

[9] Sun, Y., L. Fan and X. Hong, "Technology Development and Application of Blockchain: Current Status and Challenges," *Chinese Journal of Engineering Science* 20 (2018): 27. 10.15302/J-SSCAE-2018.02.005

[10] Wan, S., M. Li, G. Liu et al., "Recent Advances in Consensus Protocols for Blockchain: A Survey." *Wireless Network* 26 (2020): 5579–5593. 10.1007/s11276-019-02195-0

[11] Sayadi, S., S. Rejeb and Z. Choukair, "Blockchain Challenges and Security Schemes: A Survey," 2018 Seventh International Conference on Communications and Networking (ComNet), 2018, pp. 1–7. 10.1109/COMNET.2018.8621944

[12] Mingxiao, D., M. Xiaofeng, Z. Zhe, W. Xiangwei and C. Qijun, "A Review on Consensus Algorithm of Blockchain," 2017 IEEE International Conference on Systems, Man, and Cybernetics (SMC), 2017, pp. 2567–2572, 10.1109/SMC.2017.8123011

[13] Ramkumar, N., G. Sudhasadasivam and K. G. Saranya, "A survey on Different Consensus Mechanisms for the Blockchain Technology," 2020 International Conference on Communication and Signal Processing (ICCSP), 2020, pp. 0458–0464, 10.1109/ICCSP485 68.2020.9182267

[14] Mingxiao, D., M. Xiaofeng, Z. Zhe, W. Xiangwei and C. Qijun, "A review on consensus algorithm of blockchain," 2017 IEEE International Conference on Systems, Man, and Cybernetics (SMC), 2017, pp. 2567–2572, 10.1109/SMC.2017.8123011

[15] Wang, H., Z. Zheng, S. Xie, H. Dai and X. Chen, "Blockchain Challenges and Opportunities: A Survey." *International Journal of Web and Grid Services* 14 (2018): 352–375. 10.1504/IJWGS.2 018.10016848

[16] Lashkari, B. and P. Musilek, "A Comprehensive Review of Blockchain Consensus Mechanisms." *IEEE Access* 9 (2021): 43620–43652. 10.1109/ACCESS.2021.3065880

[17] Buterin, V. "Ethereum: A Next-Generation Smart Contract and Decentralized Application Platform," *Ethereum Foundation, Technical Report*, 2014. [Online]. Available: https://github. com/ethereum/wiki/wiki/White-Paper

[18] Lang, D., M. Friesen, M. Ehrlich, L. Wisniewski and J. Jasperneite, "Pursuing the Vision of Industrie 4.0: Secure Plug-and-Produce by Means of the Asset Administration Shell and Blockchain Technology," 2018 IEEE 16th International Conference on Industrial Informatics (INDIN), 2018, pp. 1092–1097, 10.1109/INDIN.2018.8471939

[19] Buterin, V. "On Public and Private Blockchains," https://blog.ethereum.org/2015/08/07/ on-public-and-private-blockchains/, 2015.

[20] Yu, G., X. Wang, K. Yu, W. Ni, J. A. Zhang and R. P. Liu, "Survey: Sharding in Blockchains." *IEEE Access* 8 (2020): 14155–14181.

3

Smart Contracts: The Self-Executing Contracts

R. Kaur, A. Ali, and Md Faisal

Glocal University, Saharanpur, U.P., India

CONTENTS

DOI: 10.1201/9781003203957-4

3.1 Introduction

A blockchain is a digital record that stores transactions publicly after nodes have verified them. The nodes validate each transaction, and the cryptography hash function secures the transactions. A transaction is linked by the hash value of the previous transaction. No one can amend or alter a transaction once it is added to the blockchain, but it can be viewed publicly, bringing transparency to the system [1]. To validate a transaction, blockchain employs some proof-of-work and proof-of-stake techniques.

In the previous years, the growth of blockchain innovation has demonstrated that it offers a vast range of use cases. The combination of blockchain technology and smart contracts allows for a great deal of flexibility to develop, design, and implement real-world problems at a lower cost and less time than traditional third-party systems [2].

This section covers introduction of smart contracts and off-chain contracts. The section also covers the literature survey.

3.1.1 Definition

The word "smart contract" was firstly given by Nick Szabo during the mid-1990s. He proposed turning contract clauses into code and embedding them in software or hardware to make the contracts self-executing, reducing contracting costs among parties, and abstain from unintentional discrepancies during the execution of the contract. This contract is a computerized code that executes on the blockchain and holds a set of protocols [3].

Generally, the term *smart contract* is used to describe the automation of legal contracts. This word describes a legal contract that could be expressed and implemented in software in whole or in part [4]. A smart contract lives on the blockchain and ensures that the required conditions fulfill the user's needs. Anyone connected to the network can see the written code, which is publicly visible in the blockchain. The contract is triggered to execute the digital transaction once the requirements are met by the specified deadline [5].

The terms of a smart contract are encrypted using cryptography; hence, no one can change the contents of a contract. The immutability offered by blockchain makes sure that a copy of the contract is stored on every node associated with the network, ensuring a backup version of the contract. The operating mechanism of a smart contract and the traditional contract between two or more parties is almost the same [6]. There is no need to rely on attorneys to create a contract for them; instead, the smart contract is automatically executed to provide payment whenever specific criteria are fulfilled (see Figures 3.1 and 3.2).

In comparison to traditional contracts, smart contracts offer the following benefits:

- Smart contracts cannot be changed arbitrarily once they have been published due to the immutability of blockchains. Furthermore, all transactions kept and mirrored across the entire distributed blockchain system may be traced and audited. As a result, nefarious activities such as financial fraud can be considerably reduced [7].

FIGURE 3.1
Regular contract.

FIGURE 3.2
Smart contract.

- Without passing via an inter-mediator, blockchains ensure the trust of the entire system through distributed consensus procedures. In a decentralized manner, smart contracts placed in blockchains can be executed automatically. This might significantly reduce the expenses of administration and services incurred as a result of third-party participation.
- Eliminating the need for a middleman can enhance the efficiency of a business process dramatically.

3.1.2 Off-Chain Contracts

These are smart contracts whose code is executed by the client rather than the miners, requesters, endorsers, or harvesters. With the help of interoperability consensus, they can:

- Conduct data-intensive computations more cost effectively
- Develop code that is not limited to a cryptocurrency incentive structure
- Write code that has less influence on the blockchain
- Develop code that is not limited to a cryptocurrency incentive structure
- Write code that has less influence on the blockchain

We want to preserve our serverless design; therefore, we'll have to rely on peer-to-peer web protocols. Let's consider how to solve the game.

As long as the game is running smoothly, that is, each player is transmitting a correct move alternately, the blockchain does not need to authenticate all of these correct moves; the clients may accomplish that. All transactions between clients are recorded on the blockchain. Each participant should be able to demonstrate the present game state independently. To prevent cheating, each report to the blockchain should include the entire game state and the other player's legitimate signature [8]. It can be cryptographically verified that both players accept that the game's current state is controlled this way [9].

In another way, during typical gameplay, each player sends their next move, the following game state, and a cryptographic signature. If the other player objects, the data can be submitted to the smart contract for verification. In the end, you can do the same thing. Clients will also acknowledge each other's communications, making it easier to find faults.

Clients only submit messages to the blockchain when they wish to claim something and to claim their prize, aside from setting up the game. The app remains serverless, secure, and untrustworthy. The blockchain is the only source of truth, but clients may help it become more efficient. However, in a problem, the game can entirely degrade and run on the blockchain [10].

Off-chain storage and computation of the smart contract to validate the state are possible. The only reason it is preferable to be on-chain is that identity generation for the off-chain smart contract on blockchain employing code signing methods protects it from being altered and maintains the immutability feature. The hash of the code, the version, and the date is all part of the code's identification. These off-chain smart contracts likewise follow the life cycle of a blockchain asset and allow for contract transfer across participants.

3.1.3 Litrature Survey

T. Hewa et al.: The study looks at some of the most important applications that have already profited from smart contracts. In addition, we discuss the future possibilities of blockchain-based smart contracts in various applications. The paper presents a comprehensive overview of blockchain-based smart contact applications. These issues should be the focus of future smart contract research. Future research opportunities include investigating consensus mechanisms, data use efficiency, reduced latency, and limited storage overheads with relatively low latency in transaction processing [11].

A. Panarello et al.: The study introduces device manipulation and data management. The paper also examine the major hurdles that the research community has faced in integrating BC and IoT smoothly, as well as the major outstanding concerns and future research paths. Finally, the research gives an overview on novel applications of BC in the machine economy [12].

S. Parjuangan and Suhardi: The goal of this research was to learn about the most recent research in the field of smart contracts. In addition, this article provided a wide overview of blockchain-based smart contract platforms. This study will aid in identifying research gaps that should be addressed in the future. The extraction results reveal that smart contract platforms are required for trade operations. This platform must be quick (in terms of producing blocks), trustworthy, secure, stable, and user-friendly [13].

C. Fan et al.: The study presents a thorough review on blockchain performance evaluation. This paper compare and contrast the existing empirical blockchain assessment approaches, such as benchmarking, monitoring, experimental analysis, and simulation, in the empirical analysis. In analytical modeling, the paper covers stochastic models that have been used to evaluate the performance of popular blockchain consensus methods. The authors extract crucial criteria for selecting the most appropriate evaluation approach for enhancing the performance of blockchain systems by contrasting, comparing, and combining diverse methodologies together [14].

D. Harz and W. Knottenbelt: The study presents a review of paradigm, type, instruction set, semantics, and metering, as well as different smart contract languages with a focus on security characteristics. The study also looks at smart contract and distributed ledger verification tools and approaches. As a result, the article describes their verification strategy, automation degree, coverage, and supported languages. Finally, future research directions are presented, including formal semantics, verified compilers, and automated verification [15].

M. Angelo and G. Salzer: The study takes into account the tools and their installation and evaluation. It is intended to serve as a guide for individuals who want to examine previously deployed code, construct secure smart contracts. The authors look into 27 tools for analyzing Ethereum smart contracts in terms of availability, maturity, methodologies used, and security risks [16].

3.2 Workings of a Smart Contract

A smart contract is a group of procedural rules and logic for scenario-response scenarios. Specifically, they are blockchain-based decentralized, trusted shared codes [17]. The parties involved must be convinced on contractual specifics, violation of contract conditions, accountability for contract breach, and external verification data sources, and then put it on the blockchain as a smart contract to automate contract execution on behalf of the signatories. There are no central agencies involved in the process (see Figure 3.3). Section 2 covers the workings of smart contracts, programming language and platforms used for Ethereum and Hyperledger Fabric, Further, this section covers the steps involved in the development of a smart contract.

3.2.1 Programming Languages and Platforms

Different blockchain systems can be employed to create smart contracts. Platforms differ in terms of their qualities [18]. We'll go through Ethereum and Hyperledger Fabric, two popular systems.

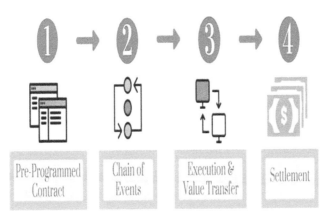

FIGURE 3.3
How a smart contract works.

3.2.1.1 Ethereum

The Ethereum platform allows applications to run precisely as they are programmed, with no downtime, censorship, theft, or interference of a third party. Ethereum applications are powered by Ether, Ethereum's platform-specific cryptographic token. Ether can be utilized as a digital currency and reward the participant nodes for their tasks [19].

3.2.1.2 Hyperledger fabric

The Linux Foundation administers this blockchain platform. This framework is used to build distributed ledger systems, and it is based on a modular design that allows for high levels of confidentiality, robustness, flexibility, and scalability [20]. It is created to allow for pluggable deployments of various components while also considering the complexities and complexities that occur throughout the economic environment.

The platform uses container technology to store smart contracts known as "chain code," which make up the system's application logic. Aside from that, "chain code" is the only way to interface with the blockchain and the only transaction data source. Smart contracts on Ethereum can be authored in several high-level languages. These codes will be compiled into EVM bytecodes and executed [21].

Go and Java languages are used to write a smart contract for Hyper ledger Fabric [22]. The key to implementing smart contracts for this platform is to use the "chain code" interface. There are three functions: Init, Invoke, and Query. These functions handle the deployment of the contract, processing of the transaction, and transaction queries, respectively.

A smart contract can be executed utilizing a unique address allocated to it by blockchain technology. For a clear understanding of how a smart contract works, consider a scenario: If someone wishes to sell a property or give it on rent to someone, they can easily leverage an existing blockchain network to deploy a smart contract [23].

Information on the property can be kept in the blockchain, and any member in the network can view it, but no one can alter it. In this manner, the owner of that property can easily find a buyer without the help of a third party [24].

Smart contracts built on the blockchain could have a lot of advantages in a variety of scenarios:

- Reliability and real-time updates
- Accuracy
- Reduced risk of execution
- There are fewer middlemen
- It is less expensive

3.2.2 Steps to Take When Developing Smart Contracts

Following are the steps to develop a smart contract [25]:

1. Recognize the smart contract's use case.
2. Generate a new architecture for smart contract interactivity, such as a flowchart showing how functions would interact.
3. Begin work with any IDE or development tool, and re-mix with detailed documentation of every function.

4. Test the smart contracts on a test-net or private blockchain after the development is complete.

5. When performing testing on the test-net, keep track of every transaction and compare the results to the exact use case of the smart contract.

6. After all this, unit testing is the next step, and there are a variety of frameworks for unit and integration testing that may be used to test smart contracts.

7. After unit testing on ganache with the truffle framework, the smart contract author should seek a third-party audit of the smart contract.

3.3 Issues Related to the Security

Section 3 covers the various security issues and the corresponding preventive measures related to smart contracts. We consider a smart contract as a digital representation of a real-world legal arrangement. Once it has been validated by the consensus process and uploaded to the blockchain, it will be executed according to the conditions of the earlier agreement without any interruption from a third party [26]. The blockchains like Fabric, Corda, and EOS support smart contracts due to their success on Ethereum. Several significant properties of today's blockchains, on the other hand, may generate security difficulties with smart contracts (refer to Table 3.1).

3.3.1 Tamper-Proofing and Decentralization

Blockchain is a decentralized, tamper-resistant digital log. Furthermore, malicious actors using pseudonyms can build and execute smart contracts. Hence, it is not easy to repair a vulnerable smart contract, and once deployed, it's easy to lose control [27].

3.3.2 Public Ledgers and Open-Source Code

A smart contract is usually open source. An adversary may be able to see contract transactions and data. Hence, smart contract vulnerabilities are easy to attack.

3.3.3 Blockchain Platforms and Smart Contract Languages

The development of d-apps on the blockchain differs from the development of regular applications. D-app developers must have a deep understanding of blockchain operations. Otherwise, smart contract creators' intentions are frequently at odds with smart contract implementation. Furthermore, because blockchain technologies advance swiftly, design faults in blockchain platforms or smart contract languages may emerge. D-app developers are constantly presented with new platform characteristics [28].

3.3.4 Transactions with Fictitious Names

Maximum transactions on public blockchains are pseudonymous. This attribute encourages illicit activity in smart contracts, such as money laundering and Ponzi scams. Financial assets are frequently transferred using smart contracts. Smart contract security

TABLE 3.1

Various Security Issues and Preventive Measures

Smart Contract Platform	Security Issues	Security Measures
Ethereum	Re-entrancy Denial of Service Integer Flow Numerical Flow	• Mark untrusted contracts • Prevent state changes after external calls • Error-Handling in external calls • Prefer pull over push for external calls • Avoid using delegate call functions to untrusted code • Manage the function code: conditions, actions, and interactions
EOS	Remote Code Execution RAM Exploit	• Authorization analysis • Evaluate the numerical overflow • Use assumptions fully in the contract • Generating true random numbers • Set the limit transfer rate
TeZos	Callback Authorization Bypass attack	• Adding owners to smart contracts • Addition of role-based access control • Prevent the storage and transfer of unsafe private data • Prevent batch operation • Ensure that a message remains unrepetitive

flaws could result in a high amount of financial losses. The DAO breach in Ethereum, for example, resulted in a hard fork of the blockchain to remove the malicious transactions. Hence, the security threat in smart contracts is more challenging than in regular contracts [29].

3.3.5 Transaction-Ordering Dependency

This problem appears when two interdependent transactions that invoke the same contract are merged into a single block. The miner chooses the order in which transactions are carried out. If the transformations are not performed in the proper sequence, an opponent can effectively launch an attack. Assume that a puzzle contract exists, with the user who completes the problem receiving a reward. When a user submits a proper puzzle solution (Tu), the evil owner immediately sends a transaction (To) to adjust the reward for the contract. By chance, those two transactions (To and Tu) might be within the same block. The user would receive a reduced payment if the miner executed To before Tu, and the wicked owner would succeed in their attack.

3.3.6 Reliance on Timestamps

The issue appears when a contract employs the timestamp of a block as a criterion to trigger and execute transactions. For example, a game-based contract that selects the winner using the timestamp as a random seed. The miner usually sets the timestamp to the current local time. The timestamp, on the other hand, has the drawback of allowing a dishonest miner to change its value by around 15 minutes while the blockchain system still accepts the block [30]. Because the timestamp value is not guaranteed to be correct, contracts that depend on it are vulnerable to attack by dishonest miners.

3.3.7 Mishandled Exception Vulnerability

This issue arises whenever a contract (caller) calls another contract (callee) without inspecting the value returned by the callee. When calling another contract, the callee contract may raise an exception. Depending on the call function's architecture, this exception may or may not notify the caller.

3.3.8 Re-entrancy Vulnerability

The problem arises when an attacker uses a recursive call function to make several repeating withdrawals. In June 2016, an attacker used the Decentralized Autonomous Organisation's (DAO) re-entrancy vulnerability to steal over $60 million.

3.3.9 Privacy Concerns

There issues concerning privacy are:

> The lack of transactional privacy is the first issue. All transactions and user balances are visible to the public in blockchain systems [31]. Because many people perceive financial transactions to be confidential information, this loss of privacy may hinder the adoption of smart contracts.

The lack of data feeds privacy is the second issue. When a contract necessitates the use of data feeds, it requests the provider of those feeds. This request, however, is public because it can be seen by anybody on the blockchain [32].

3.3.10 Performance Issues

Smart contracts are performed sequentially in blockchain systems. However, because the number of smart contracts that can be performed per second is restricted, this will severely impact the performance of blockchain systems. Blockchain solutions will be unable to scale as the number of smart contracts grows in the future.

Different security issues related to the platforms and their corresponding preventive measures are given in the following table (Table 3.1).

3.4 Applications for Smart Contracts

This section covers the different application areas of a smart contract. People in the legal, commercial, technical, and other sectors have genuine reasons to be both excited and apprehensive about the expanding prominence of smart contracts. There are numerous reasons to be optimistic. Smart contracts can improve economic efficiency, lower transaction, and legal costs, and make transactions more transparent and anonymous.

Smart contract applications or decentralized apps may be developed and tested more easily on public blockchains. Startups can use public smart contracts to raise capital. Big businesses are primarily interested in using permissioned smart contracts to incorporate their business models and enforce business rules.

Banking, electronic medical records (EMRs), and IoT data management are the most common application cases [33]. There are also some other uses, such as insurance, finance trading, e-government, etc. The following are a few examples of smart contract uses in the real world.

3.4.1 Medical Records and Health Care

Smart contracts can be effectively employed in health care and the management of access to medical records. Many healthcare practitioners view blockchain technology and smart contracts as a secure means to share and access patients' electronic medical records. Smart contracts can include multi-signature authorization among patients and providers, limiting access to the record to only authorized individuals or devices. They also make interoperability possible by using collaborative version control to keep the record consistent.

Smart contracts can also be employed to provide researchers access to some personal health-related information and enable micro-payments to be automatically delivered to patients for participation and benefit the patients and their caregivers.

3.4.2 Banking

In banking, smart contracts can be utilized to enforce laws and policies, such as in the mortgage service. A study mentions that smart contracts in mortgages might save consumers $480-960 USD per loan, while banks could save $3-11 billion USD annually in the United States and Europe. Smart contracts can also use by banks to organize clearing and settlement procedures. Additionally, smart contract logic can be used to implement KYC and anti-money laundering requirements quickly. Stellar blockchain, built on top of Hyperledger Fabric, allows for automated currency exchange in international transactions.

3.4.3 Internet of Things

The usage of smart contracts and blockchain for managing data in IoT is a potential but contentious application scenario. Intuitively, because both systems are non-centralized, blockchain might be used to improve trust in IoT ecosystems that transfer vast amounts of data regularly. To begin with, IoT data is frequently sensitive and must not be shared with others. Second, blockchains use a lot of resources. Having all IoT devices execute all applications, even with lightweight consensus procedures, is redundant due to their low processing capability. The objective is to create a reliable, cost-effective, and efficient corporate network while retaining an indelible record to meet the needs.

3.4.4 Insurance

The insurance sector chooses smart contracts to execute error checking, routing, workflow approval, and payout calculations according to the claim. For instance, the processing of travel insurance claims can automatically be checked for delayed flights or cancellations. Smart contracts can assist in removing the human element from the process, lowering total administrative costs for insurers, and boosting transparency for clients.

Nonetheless, technological restrictions and legal regulations must be solved before moving to smart contracts for insurance policies. Another disadvantage of smart contracts is their inflexibility. Traditional contracts can be altered or dissolved if all parties agree, but smart contracts, like computer programs, do not have this capability.

Each year, the expenses of insurance sector are tens of millions of dollars for claims and losses. Tens of millions of cash to bogus claims contracts that are smart can be used to automate claim processing, verification, and payment, increasing claim processing speed.

Flight insurance is being brought to smart contracts by AXA,14. If a passenger's flight is delayed by more than two hours, they will automatically be notified of the compensation alternatives. Smart contracts can also be used in the automotive industry.

3.4.5 Transactions Involving Money

Business concepts such as peer-to-peer financing and online insurance are particularly well-suited to smart contracts. Smart contracts' quickness can significantly cut transaction costs and boost efficiency, eliminating the need for time-consuming clearance and delivery. The traditional financial exchange involves collaboration by central agencies or exchanges. In contrast, smart contracts' agility can significantly cut transaction costs and improve efficiency, removing the need for time-consuming cleaning and delivery.

3.4.6 Forecasting Markets

It has been demonstrated that prediction markets can give better upcoming forecasts and speculative strategies. Smart contracts can be utilized in prediction markets because of their distributed consensus verification and immutability. Augur and Gnosis are two examples of typical uses. With the use of blockchain, Augur 8 has created a shockingly accurate forecasting tool.

3.4.7 Securities

The security industry is characterized by complex procedures that are time-consuming, wasteful, inconvenient, and risky. Smart contracts can prevent intermediaries in the securities chain custody and make automatic dividend payments, stock splits, and liability management more manageable while lowering the operational risks. Furthermore, smart contracts can help with the clearing and settlement of transactions.

3.4.8 Trade Financing

The trade finance business is now riddled with inefficiencies, making it very vulnerable. Aside from that, trading processes are still based on paper. Finance has to be improved or replaced as soon as possible. Operations that have been digitalized businesses can use smart contracts to save time and money. To automatically initiate promotional actions depending on criteria established criteria that will increase efficiency by reducing operations and lower fraud and compliance risks.

3.4.9 Organizational Management

In today's world, most businesses are managed by their executives. They are governed by and centered on a board of directors who make the majority of the decisions. It is expected that the future management of the firm will be decentralized and flattened smart contracts can eliminate unneeded intermediaries who impose arbitrary constraints, limitations, and rules that are unduly complicated.

3.4.10 E-Government

Smart contracts can boost efficiency and authority by simplifying bureaucratic processes. For instance, China built the first e-government service platform employing blockchain and smart contracts in Foshan, China. Smart contracts can also be employed in other areas of e-government. E-voting creates innovative payment mechanisms for employment and retirement, enhancing international aid systems, etc.

3.5 Challenges

Although smart contracts are a promising technology, there are still a lot of obstacles to overcome. This section provides a summary of recent progress in addressing these issues [34].

3.5.1 Creation-Related Challenges

The establishment of contracts is a crucial stage in the implementation of smart contracts. Users must create their contracts, which they must deploy on various blockchain platforms because blockchains are fundamentally immutable; blockchains can't change smart contracts built on them after they've been implemented [35]. Hence, developers must carefully handle the issues listed below.

Programing languages such as Go, Kotlin, and Java are used to develop smart contracts. After that, the source code will be compiled and run. Hence, programs have different types of codes depending on the period. This is still an important issue to make programs readable in every form.

3.5.2 Functionality-Related Challenges

The existing platforms for smart contracts face numerous functional challenges. Following are some examples:

- Re-entrancy: It refers to the ability to recall an interrupted function safely. Malicious individuals could use this flaw to steal digital currency.
- Eliminate randomization: Randomization of created blocks may be necessary for some applications, like lotteries and others. This can be performed by generating pseudo-random integers using a timestamp or nonce. On the other side, a few threatful miners may create blocks to deviate from the pseudo-random generator's findings.
- Excessive billing: According to new research, smart contracts can be overpriced as a result of under-optimization [36]. These overloaded patterns are characterized by dead code and expensive operations in loops, including repeating computations.

3.5.3 Deployment-Related Challenges

Smart contracts are being deployed to the well-suited platforms after their creation. Additionally, to avoid any potential hurdle, testing of the smart contracts is compulsory [37]. The creators of a smart contract should also consider the contract's interaction patterns so that any potential losses due to malicious activity can be avoided.

It is practically impossible to make modifications to smart contracts once they have been put on blockchains. Hence, it is critical to evaluate the accuracy of smart contracts before their official deployment. Moreover, because of the complexity of modeling smart contracts, verifying their accuracy might be difficult [38].

3.5.4 Execution-Related Challenges

The execution step is critical for smart contracts since it defines their final state. While executing smart contracts, a variety of challenges must be handled. One cannot execute smart contracts without accessing real-world data [39–43]. A smart contract is designed so that it requires running in a sandbox. An oracle acts as an agent in a smart contract, locating and verifying real-world events and relaying this information to the smart contract. Hence, finding a reliable oracle becomes a task.

3.6 Conclusion

Because of the swift advancement of blockchain innovation, smart contracts have emerged as a popular study topic. Smart contracts' unchangeable and irreversible qualities can allow people to exchange money, shares, and intellectual assets with no interference from a third party. As a result, smart contracts are projected to disrupt several established industries, including finance, health care, business, and many more. This chapter provides an overview of smart contracts, covering their working, platforms, security issues, and application scenarios. We also went through some of the smart contract's challenges and future trends.

References

[1] di Angelo, M. and G. Salzer, "A Survey of Tools for Analyzing Ethereum Smart Contracts," 2019 IEEE International Conference on Decentralized Applications and Infrastructures (DAPPCON), 2019, pp. 69–78.

[2] Singha, A. et al., "Blockchain Smart Contracts Formalization: Approaches and Challenges to Address Vulnerabilities." *Computers & Security* 88 (2020): 101654.

[3] Wang, S. et al., "Blockchain-Enabled Smart Contracts: Architecture, Applications, and Future Trends." *IEEE Transactions On Systems, Man, And Cybernetics: Systems* 49 (2019) (11): 2266–2277.

[4] Huang, Y. et al., "Smart Contract Security: A Software Lifecycle Perspective." *IEEE Access* 7 (2019): 150184–150202.

[5] Alharby, M. and A. Van, Moorsel. Blockchain-based smart contracts: A systematic mapping study. arXiv preprint arXiv:1710.06372, 2017.

[6] Amani, S., M. Begel, M. Bortin and M. Staples, "Towards Verifying 'Ethereum Smart Contract Bytecode in Isabelle/hol," Proceedings of the 7th ACM SIGPLAN International Conference on Certified Programs and Proofs, 2018, pp. 66–77.

[7] Assal, H. and S. Chiasson, "Security in the Software Development Lifecycle," In Proceedings of the 14th Symposium on Usable Privacy and Security (SOUPS' 18), 2018, pp. 281–296.

[8] Assal, H. and S. Chiasson, "'Think Secure From The Beginning' a Survey With Software Developers," In Proceedings of the 2019 CHI Conference on Human Factors in Computing Systems, 2019, pp. 1–13.

[9] Chang, J., B. Gao, H. Xiao, J. Sun, Y. Cai and Z. Yang, "Compile Critical Path Identification and Analysis for Smart Contracts," In Proceedings of the International Conference on Formal Engineering Methods, 2019, pp. 286–304.

[10] Chen, H., M. Pendleton, L. Njilla and S. Xu, "A Survey on Ethereum Systems Security: Vulnerabilities, Attacks, and Defenses." *ACM Computing Surveys (CSUR)*, 53 (2020) (3): 1–43.

[11] Hewa, T. et al., "Survey on Blockchain Based Smart Contracts: Applications, Opportunities and Challenges." *Journal of Network and Computer Applications* 177 (2021), Article Number: 102857. 10.1016/j.jnca.2020.102857

[12] Panarello, A. et al., "Blockchain and IoT Integration: A Systematic Survey." *Sensors* 18 (2018) (2575). 10.3390/s18082575

[13] Parjuangan, S. and Suhardi, "Systematic Literature Review of Blockchain based Smart Contracts Platforms," International Conference on Information Technology Systems and Innovation (ICITSI), 2020, pp. 381–386. 10.1109/ICITSI50517.2020.9264908

[14] Fan, C. et al., "Performance Evaluation of Blockchain Systems: A Systematic Survey," *IEEE Access*, 8 (2020): 126927–126950, 10.1109/ACCESS.2020.3006078

[15] Harz, D. and W. Knottenbelt, Towards Safer Smart Contracts: A Survey of Languages and Verification Methods, arXiv:1809.09805v4 [cs.CR] 1 Nov 2018.

[16] Angelo, M. and G. Salzer, "A survey of tools for analyzing Ethereum smart contracts," IEEE International Conference on Decentralized Applications and Infrastructures (DAPPCON), IEEE Computer Society, 2019, pp. 69–78. 10.1109/DAPPCON.2019.00018

[17] Kaur, R., A. Ali, and A. Novel, "Blockchain Model for Securing IoT Based Data Transmission," *International Journal of Grid and Distributed Computing* 14 (2021) (1): 1045–1055.

[18] Zhou, Y., D. Kumar, S. Bakshi, J. Mason, A. Miller and M. Bailey, "Essays: Reverse Engineering Ethereum's Opaque Smart Contracts," In Proceedings of the 27th USENIX Security Symposium (USENIX Security' 18), 2018, pp. 1371–1385.

[19] Yamashita, K., Y. Nomura, E. Zhou, B. Pi and S. Jun, "Potential Risks of Hyper Ledger Fabric Smart Contracts," In Proceedings of the IEEE International Workshop on Blockchain Oriented Software Engineering (IWBOSE' 19), 2019, pp. 1–10.

[20] Suiche, M., "Porosity: A Decompiler for Blockchain-Based Smart Contracts Bytecode," DEF CON, 2017, pp. 1–30.

[21] Rodler, M., W. Li, G. O. Karame And L. Davi, "Serum: Protecting Existing Smart Contracts Against re-Entrancy Attacks," In Proceedings of the Network and Distributed Systems Security Symposium (NDSS '19), 2019.

[22] Kalra, S., S. Goel, M. Dhawan and S. Sharma, "Zeus: Analyzing Safety of Smart Contracts," In Proceedings of the Network and Distributed Systems Security Symposium (NDSS '18), 2018, pp. 1–15.

[23] Hirai, Y., "Defining the Ethereum Virtual Machine for Interactive Theorem Provers," In Proceedings of the International Conference on Financial Cryptography and Data Security, 2017, pp. 520–535. Springer.

[24] Grishchenko, M. M. and C. Schneidewind, "A Semantic Framework for the Security Analysis of Ethereum Smart Contracts," In Proceedings of the International Conference on Principles of Security and Trust, 2018, pp. 243–269. Springer.

[25] Sharma, A. et al., "Blockchain Based Smart Contracts for Internet of Medical Things in e-Healthcare." *Electronics* 9 (2020): 1609.

[26] Zou, W., "Smart Contract Development: Challenges and Opportunities." *IEEE Transactions on Software Engineering* 47 (2019) (10): 2084–2106. doi: 10.1109/TSE.2019.2942301

[27] Sayeed, S. et al., "Smart Contract: Attacks and Protections." *IEEE ACCESS* 8 (2020): 24416–24427.

[28] Mohanta, B. K. et al., "An Overview of Smart Contract and use Cases in Blockchain Technology," 9th ICCCNT IISC, Bengaluru Bengaluru, India, July 10–12, 2018.

[29] Luu, L., et al., "Making Smart Contracts Smarter," ACM SIGSAC Conf. Comput. Commun. Security.-CCS, New York, NY, USA, 2016, pp. 254–269.

[30] Brent, L. et al., Vandal: A scalable security analysis framework for smart contracts, arXiv:1809.03981. [Online], Sep. 2018.

[31] S. Kalra et al., "ZEUS: Analyzing Safety of Smart Contracts," In Proceeding Network Distribution System Security Symposium, 2018, pp. 1–12.

[32] Jarzabek, S. et al., *Integrating Research and Practice in Software Engineering Cham*. Switzerland: Springer, 2019.

[33] Bamako, S., Detecting Critical Smart Contract Vulnerabilities with re: MythX. Accessed: Oct. 19, 2019.

[34] Christidis, K. and M. Devetsiokiotis, "Blockchains and Smart Contracts for the IoT," *IEEE ACCESS* 4 (2016): 2292–2303.

[35] Sariboz, E. et al., Off-chain Execution and Verification of Computationally Intensive Smart Contracts, IEEE 2021, arXiv:2104.09569v3 [cs.CR] 25 Apr 2021.

[36] Bartoletti, M. et al., "An Empirical Analysis of Smart Contracts: Platforms, Applications, and Design Patterns," In Proceedings of the International Conference on Financial Cryptography and Data Security, Sliema, Malta, 3–7 April 2017, pp. 494–509.

[37] Szabo, N., "Smart Contracts: Building Blocks for Digital Markets," Available Online: https://kameir.com/smart-contracts/

[38] Sillaber, C. and B. Waltl, "Life Cycle of Smart Contracts in Blockchain Ecosystems." *Datenschutz und Datensicherheit – DuD Springer* 41 (2017) (8): 497–500.

[39] Khatoon, A. "A Blockchain-Based Smart Contract System for Healthcare Management," *Electronics* 91 (2020): 94.

[40] Li, C. et al., "Scalable and Privacy-Preserving Design ofOn/Off-chain Smart Contracts," 35th International Conference on Data Engineering Workshops (ICDEW) IEEE, 2019, pp. 7–12.

[41] Ali, A. and A Concise, "Artificial Neural Network in Data Mining." *International Journal of Research in Engineering and Applied Sciences* 2 (2012) (2): 418–428.

[42] Parveen, N., A. Ali and A. Ali, "IOT Based Automatic Vehicle Accident Alert System," IEEE 5th International Conference on Computing Communication and Automation, 2020, pp. 330–333.

[43] Sachdeva, S. and A. Ali, "A Hybrid Approach Using Digital Forensics for Attack Detection in a Cloud Network Environment." *International Journal of Future Generation Communication and Networking* 14 (2021) (1): 1536–1546.

4

Introduction to Internet of Things with Flavor of Blockchain Technology

Waqas Ahmed Siddique[1], Awais Khan Jumani[1], and Asif Ali Laghari[2]

[1]Ilma University Karachi, Sindh, Pakistan
[2]Sindh Madressatul Islam University, Karachi, Sindh,
 Pakistan

CONTENTS

DOI: 10.1201/9781003203957-5

4.1 Introduction

Smartphones play an important part in the IoT, owing to the fact that many IoT applications can be managed via an application (app). In order to provide the optimal condition for you when you get back from a job, you may interact with your intelligent thermostat using your smartphone, for example. What is another plus? Using this method, you may avoid using unnecessary air conditioning while you are away, possibly reducing you funds on energy expenses. IoT gadgets include detectors and mini-computer units that, through machine learning, react on the information collected by the instruments. In essence, IoT gadgets are tiny computers that are linked to the Internet and are thus susceptible to viruses and hacking. When computers understand in a manner similar to that of people by gathering data through their surroundings. This is known as machine training and it is what allows IoT devices to become smart. This information may be used to assist the computer in learning your preferences and adjusting itself appropriately. Machine learning is a form of artificial intelligence that allows machines to understand on their own, without the need for human intervention.

That does not imply that your smart speaker will engage in a conversation with you about the most important aspects of yesterday night's big game. However, if your linked refrigerator detects that you are close to a supermarket, it may give you a warning on your cell phone that you are short on milk and eggs, prompting you to replenish your supplies [1].

This IoT is intended to make the process easier in a variety of ways. Here seem to be a few illustrations:

1. A smart bathroom measures that operate in conjunction with your running to send meal preparation suggestions to your computer or desktop, allowing you to maintain your fitness goals.

2. You may have your house monitored by security equipment that will automatically switch the lights up and down as you approach and leave rooms and record live so you can look around while you are absent.

3. Using intelligent voice commands, you can place your regular takeaway order on demand, making it a snap to have freshly prepared food served to your doorstep.

4.1.1 Home Security

The IoT is the primary driving force beside a fully intelligent and protected home environment. In order to offer 24-hour/365 security, the IoT links a wide range of sensors, bells, cameras, lights, and mics that can all be managed from a mobile device. Using a PC, laptop, or smartphone, for example, the Ring doorbell video security system enables users to see, hear, and talk to guests at their front door.

4.1.2 Activity Tracker

Wearing these monitor gadgets across the day allows them to evaluate and communicate important health indications in real time, such as tiredness, hunger, physical mobility, respiration in blood, blood pressure, surveillance systems, and compliance with medication regimens, among others. A decrease in the frequency of urgent doctor or hospitalizations and an increase in the independence of senior or handicapped individuals are all benefits of at-home health screening. Figure 4.1 show a model for an indoor security system.

4.1.3 Digital Twins

In the production industry, a digital dual is basically a digital duplicate of a physical element that is comparable to the real counterpart. The digital duplicate may be updated in real time when the physical item alters in reaction to its environment, thanks to advancements such as the Internet of Things, artificial intelligence, and machine learning, among many others. Engineers may make changes or verify upgrades on the digital twin rather of the actual physical product, saving time and money.

FIGURE 4.1
Best indoor security system [2].

4.1.4 Self-Healing Machine

A manufacturing apparatus may be engineered to detect and rectify variations in its own functioning based on arrays of numbers of detectors, artificial intelligence, and machine learning. This will prevent issues from arising that would need downtime and maintenance. Organizations save both time and expense as a result of this, and workers who might usually monitor machinery and perform maintenance may devote their time to relatively high activities.

4.1.5 AR Glasses

Google Glass is basically a compact, lightweight gadget that will be like wearing a pair of goggles and allows users to operate without having to use their hands. In the glasses' "lenses," the information is showcased in a way that enables customers to access a range of Internet apps, such as Google Maps and Gmail. Workers in industrial environments often utilize them to bring up designs or product specs, among other things. The most recent version of these smart glasses enables employees to broadcast clear "point of view" video from professional assistants in distant places via Wi-Fi, allowing them to collaborate more effectively.

4.1.6 Ingestible Sensor

To supervise medical conditions and transmit information from within the gastro-intestinal tract, ingestible electronic devices the size of a pill and equipped with a power source, microprocessor, control system, and detectors can be consumed. For example, they can be used to detect loss of blood or the absorption of pharmaceutical products. Abilify MyCite was the first pill to be authorized by the U.S. Food and Drug Administration (FDA) because it had a digital monitoring device. Figure 4.2 shows the product of Google known as the Google Glass.

4.1.7 Smart Farming

Nowadays, farming may be a very technologically advanced undertaking. Farmers are increasingly relying on IoT-enabled equipment to monitor weather, soil composition, soil moisture levels, crop health and growth, as well as animal activity and movement. Additionally, the data may be used to identify the optimum time to harvest plants, as well as to develop fertilizer profiles and scheduling plans. Drones may also be used to gather images and data about the environment.

FIGURE 4.2
Google Glass smart wearable [3].

FIGURE 4.3
Industrial security and safety [4].

4.1.8 Smart Contact Lenses

Smart contact lenses, which can gather patient information and address certain vision issues, are now the subject of extensive research in the United States. The Triggerfish intelligent contact lens, invented by the Swiss manufacturer Sensimed, is an invasive intelligent contact lens that involves a detector implanted in a smooth silicone contact lens that detects small changes in an eye's volume, which can be an indication of glaucoma. Using wireless technology, the gadget sends data from the sensor to an adhesive antenna wrapped all around eye.

4.1.9 Industrial Security and Safety

Figure 4.3 shows the model used for the purpose of industrial security and safety. It is possible to employ sensing devices and cameras to monitor the perimeter of prohibited areas and to identify intruders in regions that have not granted permission. Harmful chemicals leaks and stress buildups may be recognized and rectified immediately on, before they become significant issues. Locating and repairing fluid leaks helps to prevent damage and keep maintenance expenses to a minimum. Detection systems powered by the IoT and used to track chemical plants, nuclear power plants, and mining activities.

4.1.10 Motion Detection

Small instabilities and irregularities that may lead to catastrophic failures in vast structures such as constructions such as buildings, bridges, and dams can be recorded and analyzed by using algorithms to control motion or vibration in these infrastructures. Landslide, avalanche, and earthquake detector connections are also deployed in regions that are vulnerable to these disasters.

4.1.11 Unlimited Possibilities

The IoT is still in the early stages of development. In a few years, we will indeed be linked in manners that are impossible to fathom now, let alone today. Applications based on the IoT, particularly when coupled with automation and artificial understanding, will enhance judgment capabilities, sustainability, flexibility, wellness, and power efficiency. Through the fusion of these platforms, innovative thinking and new solutions across a broad variety of sectors will be facilitated as an outcome.

4.2 Generation of IoT Life Cycle

The IoT has a relatively straightforward development life cycle. Checking, maintaining, and controlling are all handled after deployment, and regular upgrades and disassembly

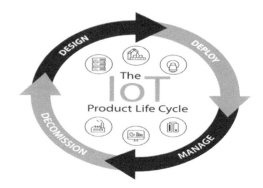

FIGURE 4.4
IoT device life cycle [5].

are performed at the conclusion. The IoT product life cycle is described in the diagram (Figure 4.4).

There are certain benefits and drawbacks to IoT devices, aside from the information presented previously, and these could have a significant effect on the existing and up-coming civilizations of humanity.

4.2.1 Advantages of IoT Devices

There are several advantages of these smart devices and some of them are given below [6]:

1. The IoT promotes device-to-device communication, also known as machine-to-machine communication.
2. It has a high degree of automation and supervision.
3. It is better to manage since it is integrated with more tech capabilities.
4. IoT has a robust monitoring capability.
5. It helps you to save huge amount of time.
6. The IoT helps to reduce the cost by eliminating manual workload and activities.
7. Streamlining everyday activities leads to better device monitoring.
8. Efficiency gains and time savings have increased.
9. Good factors contributed to a higher standard of living.

4.2.2 Disadvantages of IoT Devices

There are a number of benefits, but there are also some drawbacks.
Enlisted below are the various demerits:

1. There is no worldwide interoperability standard for IoT devices.
2. They have the potential to grow too complicated, ending in failure.
3. A compromise of confidentiality or protection may damage IoT equipment.
4. Users' safety is jeopardized.
5. Reduction in overall use of physical labor, which results in loss of jobs.
6. With the advancement of AI technology, the IoT gadget may eventually gain charge of someone's life.

4.2.3 Features of IOT

It is not a work of science fantasy. With the World Wide Web, gadgets that understand our choices and deliver the encounters we want, we are living linked lives that are more comfortable than ever before. Furthermore, the technology that allows us to link our lives is becoming more advanced. In the 1980s, students at Carnegie Mellon University created the world's earliest Internet-connected gadget, which became a worldwide phenomenon. This particular Coke vending machine would inform the developers if the beverage was cold sufficient for them to be interested in making the journey from their workstations to the station. Ever since, the IoT has expanded. What will happen in the future? Devices are becoming smaller and more intelligent. The Internet may one day be accessible from everywhere, at any time, from everything from your toothbrush to your toaster. Your devices will transform into home assistants, each communicating with the others and cooperating to provide you with greater service.

IoT security and privacy: As with every technology, there are both positive and negative aspects to consider. Convenience is a positive thing. However, being connected may also provide an entry point for hackers. Cyberattacks already get access to information such as bank logins, credit card details, and other personal information. A high level of security is required. Consider some of the possible dangers associated with the IoT and those tiny CPUs. Even if someone gains access to your IoT security cameras and begins to monitor your every move? What happens if a cybercriminal gains access to your smart TV, smoke alarms, or front-door lock and takes control of them? Vulnerabilities may arise because of being connected. Making ensuring these interconnections are secure and safe is wise; in the same way, that you shield yourself from cybercriminal activity is wise. The IoT will continue to provide new possibilities for cybercriminals to launch attacks online. One of the most effective methods to assist in the defense of your networked devices is to ensure that your gateway is protected. In this manner, it may secure your home Wi-Fi network as well as the devices that are linked to it. Consider your Wi-Fi router to be the gateway that leads into your virtual environment [7].

The IoT has been quickly changing the way we conduct business and go about our everyday lives, and it will continue to do so. Sensor technology, as well as real-time data collection and processing, may be used to monitor almost every element of our activities. Data from the IoT may be used to simplify corporate operations and increase productivity, as well as enhance our health and safety. It can also be utilized to automate activities and enable us to take a more in-depth look at our connections, systems, and surroundings.

4.3 IoT Work with Blockchain

IoT allows Internet-connected devices to transmit data to private blockchain networks, which generate hack records of shared transactions. IBM blockchain allows you to exchange and retrieve IoT data with your business partners without the requirement for centralized control and administration. To avoid conflicts and establish confidence among all permissioned network users, each transaction may be validated.

4.3.1 Benifits of IoT and Blockchain

Following are some of the benefits of the blockchains.

4.3.2 Build Trust in Your IoT Data

Every transaction is logged, stored in a separate data cluster, and linked to a protected, unchangeable over time data chain that can only be updated to.

4.3.3 Relay On Added Security

You may choose which data to administer, investigate, modify, and share with highly secure customers and partners.

4.3.4 Gain Greater Flexibility

The IBM Blockchain Platform is open, compatible, and designed for multi-cloud environments, based on the most recent edition of the premier Hyperledger Fabric platform, which is tailored for Red Hat OpenShift.

4.3.5 Generate New Efficiencies

IBM blockchain uses data from IoT devices and sensors to simplify operations and generate new business benefits throughout your network.

4.4 Application of IoT and Blockchain

Shifting shipments is a complicated operation that involves a number of stakeholders with varying levels of importance. The degrees, location, time of arrive, and condition of cargo ships may all be recorded on a blockchain that is permitted by the IoT. Transfers on the blockchain are immutable, which helps to guarantee that all stakeholders can rely on the facts and act swiftly to transfer goods rapidly and effectively.

4.4.1 Component Tracking and Complaince

Controlling and tracking all of the parts and equipment that move into a plane, car, or other item is essential for both security and regulation accountability reasons. The storage of IoT material in common blockchain ledger accounts and allows all stakeholders to track the origin of components throughout most of the life cycle of a unit. Distributing this data with governing authorities, transporters, and makers is a simple and cost-effective process that is both secure and simple.

4.4.2 Log Operational Maintenance Data

The IoT gadgets monitor the condition of security for essential equipment and the status of their servicing. From motors to lifts, blockchain offers a secure, tamper-proof record of operating data and the upkeep that results from this data. In order to do preventative repair, third-party restoration providers may observe the blockchain and then report their activity back on the blockchain. Operational records may also be communicated with regulatory organizations in order to ensure that they are in accordance.

4.4.3 Internet of Things Realization in 5G Data

Over the past 40 years, the technology has grown from basic peer-to-peer connections to a sophisticated IoT environment (Figure 4.5). Everything surrounding us is growing smart as a result of the widespread use of IoT devices. The ability to link individuals and objects at any moment from any location with internet connectivity, to get knowledge about the thing, to control the item, or perhaps both, is now available [8] and this has the potential to make the lives of humans simpler. The number of linked devices continues to grow at an exponential pace. It is predicted that the quantity of linked devices would reach 27 billion by the year 2024 [9]. The variety of these IoT software solutions will likewise increase, ranging from basic home appliances to expedition medical systems [10]. These application situations need a wide range of performance criteria, including low delay, incredibly, security systems, and large data throughput, among others.

The majority of these linked gadgets will interact with one another via wireless connections. The current wireless architecture, on the other hand, is incapable of keeping up with the fast increase IoT communications, as well as traditional mobile links. Furthermore, such infrastructures are unable of meeting the varied quality of service (QoS) needs of various IoT application situations. As a response, the 5G infrastructure is being developed with the assistance of a variety of innovations in order to meet these needs.

4.4.4 Improving Security

In the IoT, safety is a basic need. In today's world, billions of asset gadgets in heterogeneous IoT programs are being integrated into people's daily routines and activities. Intruders are increasingly targeting IoT apps because they exchange confidential material or are engaged in security activities [11].

1. **Detailed description of the restriction:** The majority of IoT devices are designed with resource limitations in mind. As a result, implementing information security on IoT devices is challenging. Device manufacturers prefer to utilize standard credentials IoT s equipment, and the vast majority of device users are not

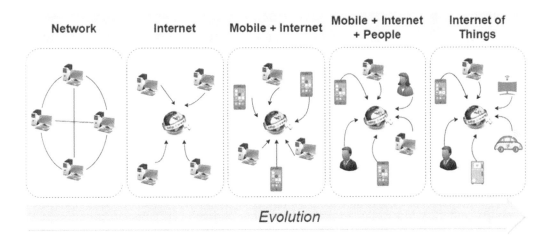

FIGURE 4.5
Evolution of IoT network [8].

interested in changing these basic passwords. Consequently, while being utilized by users, the IoT devices become more susceptible to keyword threats [12]. This will aid enemies in gaining control of the IoT devices. A variety of various kinds of assaults in the IoTs, including distributed denial of services (DDoS), man in the middle (MITM), zero-day exploits, IoT malware attacks, and malware [13], may arise because of these possible causes. Network protection methods are essential for mitigating these types of assaults. The security criteria for different IoTs applications are different. Consider the fact that the security needs for healthcare monitoring applications are very different from those for an intelligent grid program. To meet the varied security needs of any IoT application running on a telecommunication channel, separate IoT security methods must be implemented in each telecommunication topology.

2. **Network Slicing Offers a Wide Range of Protection Options:** Network chopping may be used to create a wide range of security alternatives for a wide range of IoTs applications [14]. Slice separation aids in minimizing the effect of security threats as well as safeguarding sensitive data gathered by IoT devices, among other things. The dynamic installation of protection NFs in the segments enables the detection and mitigation of threats in real time without interfering with the operation of other IoT services. The ability to construct quarantine slices in order to separate gadgets that exhibit suspicious behavior makes it easier to run such devices under more stringent conditions until the required measures can be undertaken. As a result, network layer may be helpful in terms of enhancing the security of IoT applications. Table III summarizes the types of attacks that may occur in IoT applications, as well as the types of solutions that could be given by network slicing in order to prevent such attacks. They have suggested techniques to combat IoT-based DDoS assaults in [15] and [16], which are slice separation and edge computing, accordingly, to combat the attacks.

4.5 Network Slicing Based IoT Applications

The IoT has implemented in a variety of implementation areas. As per their needs, these IoT programs may divided into four major categories. As a result, the emphasis of this part is on addressing important IoT use scenarios and applications, as well, as how network splitting may utilized to solve the difficulties encountered in these applications.

4.5.1 Smart Transportation

With the widespread evolvement of the transport network, the IoT has become closely connected with a variety of regions in shipping, including vehicle-to-vehicle interaction, vehicle-to-infrastructure interaction, independent or semi-autonomous driving, and in-car infotainment technologies. In vehicle-to-everything (V2X) systems, ultra-reliability and very low delay are key networking demands [17]. Specifically, the 3GPP distinguishes four distinct kinds of V2X interaction methods, which are as follows: vehicle-to-vehicle (V2V), vehicle-to-infrastructure (V2I), vehicle-to-pedestrian (V2P), and vehicle-to-network (V2N).

4.5.2 Inustrial Automation/Industrial IoT (IIoT)

The 4th Corporate Transformation (Industry 4.0) [18], a new improvement in industrial control systems, integrates contemporary networking and computation techniques, such as cloud computing and the IoT, to industrial production processes for the first time [19]. The outcome will be the interconnection of a vast number of IoT devices, equipment, and services with diverse network needs, allowing for the realization of Industry 4.0 [20]. In a manufacturing line, using linked IoT-based equipment rather of manual effort, as well as remote control and reporting, would improve the productivity of the network while decreasing the consumption of the service overall. The factory may be equipped with a variety of IoT devices that can be deployed around the facility to detect environmental factors and machine performance in order to reduce unexpected unavailability.

4.5.3 Smart Health Care

As with the other service domains, the IoT is extensively utilized in the medical industry, with a variety of application examples [21]. Current progress in the IoT have been highlighted as a possible alternative to relieve constraints on healthcare organizations, such as a lack of medical personnel and higher health expenditures, while still providing high-quality treatment to individuals [22]. In the healthcare industry, one of the most quest IoT healthcare technologies is virtual surgeries, which allowed users to get the services of a group of specialists who are located in various locations across the globe [23]. Remote supervision of seniors and sick that need constant supervision is an IoT usage that necessitates the use of networking solutions that are ultra-reliable and low latency. As with the rapid advancements in the area of robotics, the placement of robotic arms in close proximity to senior citizens will enable them to continue to live freely at their residence. Wearable IoT gadgets are becoming more common in a variety of intrusion detection situations, including sugar, ECG, BP, and heart rate, with the goal of preventing avoidable fatalities [24,25].

4.5.4 Smart Home and City

Because of this expansion, smart homes are now being considered for use in smart workplaces and structures and eventually intelligent buildings. In home automation and city apps, the IoT may make people's lives better by organizing day-to-day tasks. Other benefits include cost savings and energy conservation. Simple heat sensor systems that automate the operation of the air conditioning system to sophisticated picture processing systems that identify intruders and alert the authorities to enhance home security are all possible with smart home technologies. While IoT devices provide many benefits from a home automation standpoint, the pace at which consumers integrate IoT devices into their homes is dependent on the willingness of the users themselves to purchase the devices. In order to make their choice, safety and efficiency have been recognized as the most important considerations [24]. Because of rapid population expansion and urbanization in metropolitan areas, most of the world's resources are being used up in a relatively small amount of space on Earth. The rapid growth of cities must handled in a sustainable manner while simultaneously improving the overall quality of life for the population. Using the smart city idea, the IoT has been used to address these problems [26].

IoT academic and enterprise are becoming more interested in this rapidly developing field. It is clear that IoT platforms are maturing, as shown by the growing variety of IoT solutions in a variety of sectors, spanning from home automation to manufacturing systems and business sector, which includes the so-called industrial IoT. Even after this accomplishment, though, there are still certain problems to address. The most significant of these concerns, and one which is most likely to impede the uptake of IoT, is safety. Designing unsafe models, structures, and installations is made possible by the heterogeneity of procedures, software platforms, and gadgets, as well as the lack of widespread acceptance of sample solution. Moreover, IoT devices are often linked with critical material, key systems, and resources, making them particularly vulnerable to security breaches, denial of service assaults, and other forms of cybercrime and criminal activity. Because of the limited abilities of IoT equipment and the decentralized structure of IoT designs, traditional security methods are often found to be inapplicable none the IoT. Blockchain is a system that is presently getting a lot of attention, and it has the potential to be useful in ensuring security in IoT situations. With its decentralized design and capacity to offer data immutability and non-repudiation services, blockchain technology seems to be a viable method for safeguarding the IoT and preserving user/data confidentiality. The special issue on "Blockchain Confidentiality and Protection for the IoT" attempts to investigate the cutting-edge improvements, innovations, and obstacles linked to blockchain, safety, and confidentiality for the IoT that have emerged from both recent research operations and current projects, as well as to provide a forum for discussion. The presented topic is categorized by numerous open issues that need to be resolved or bettered, and for this reason, despite the large number of documents that were obtained, only 15 authentic and high-quality writings were chosen for inclusion in this unique issue of the *Journal of the American Chemical Society*. Each article was examined by a number of experts and went through many stages of the peer-review method before being accepted for publication. The current assessment of blockchain technologies and the application of blockchain innovations to e-health privacy administration were both presented in two fascinating review articles [27], which we found to be excellent. The first, in particular, is to provide a comprehensive assessment of the existing blockchain evaluations methods, as well as to highlight the associated difficulties and limits that exist in terms of their application. The writers outline the primary metrics associated with blockchain assessment and suggest an appealing design and control classification based on a critical review of the literature, while also recognizing the fully existing difficulties as well as future prospects and development in the field of blockchain technology. Rather than that, this article covers the latest developments in decentralized identity administration utilizing blockchain technology in order to identify potential possibilities for using decentralized identity administration methods for future health identification systems. The authors present a fine-grained data sharing system that is both safe and lightweight for use in a mobile cloud computing environment. With this project, the goal was to move the bulk of time-consuming activities away from resource-constrained mobile devices and onto the cloud instead [28]. It is possible to achieve a CCA security level by introducing the following novelties:

1. Supporting provable outsourced decoding, i.e. the mobile user can make sure the legitimacy of the reshaped cipher text returned from the cloud server;

2. Outsourcing decoding for intensive computing tasks during the decryption phase without disclosing user data or the decrypted key; and

3. Introducing the novelties described above. The classification demonstration and efficiency study demonstrate that the new method is both safe and appropriate for portable cloud computing settings, as shown in the paper.

In order to improve the variables of the hybrid model, the authors use advanced deep learning (ADL) methods, such as long short-term memory (LSTM) and gated recurrent units (GRU), in conjunction with a genetic algorithm (GA) to develop a forecast model [29]. They use GA to determine the optimum training variables, and then they cascade LSTM with GRU to get the desired result. The manuscript's primary goal was to assist supply chain professionals in taking use of cutting-edge technology, as well as to assist the industry in formulating regulations in accordance with ADL's forecasts for the future. An architectural framework for the IoT in order to enable identification and ensure data integrity the outlined approach discusses the security by design philosophy by mixing some of the innovations such as secure multi-party computation (SMPC) for grounded rules and distributed ledger technology (DLT) for an immutable and clear registration system. Secure multi-party computation (SMPC) is used for grounded policy rules and distributed ledger technology (DLT) is used for an immutable and transparent registry [30]. To address the complex ecosystem of intelligent movement and transportation systems, the researchers in blockchain security and privacy for the IoT propose a blockchain-based design as a trust reference network that protects user privacy while also providing reliable services to consumers. Also suitable with the legacy intelligent transportation system (ITS) architecture and offerings is the ITS-X platform. Furthermore, the hierarchical structure of chains facilitates the scalability of the system, while the use of smart contracts allows for the introduction of new services into the ITS in a more flexible manner than previously possible [31]. By using an Ethereum proof of concept implementation, the suggested architecture is shown, and the results of the depicted tests confirm the viability of the proposed design. Using smart contracts, the authors suggest a new blockchain-based system for observing patient vital signs that would be used in conjunction with other systems. For the design and development of this system, the hyperledger fabric platform was used [32]. Hyperledger fabric is a corporation ledger architecture that is used for the development of blockchain-based apps. Patients will benefit from the proposed method in a number of ways, including a comprehensive, immutable background record and worldwide access to medical records from any location at any time. When acquiring physiological data, the Libelium eHealth toolkit is being used, and the achievement has been measured in terms of transactions per second, transaction latency, and resource usage using a selected classification tool known as Hyperledger Caliper, demonstrating how the current proposal outperforms the conventional health care system when monitoring patient information. The authors offer an IoT security transmission and storage solution for sensing pictures for blockchain applications. Using intelligent sensing, the suggested method detects and categorizes user picture information, which is then divided into intelligent blocks. By using sophisticated encryption techniques, different chunks of data may be secured and sent safely over the internet. At the conclusion of the day, a sophisticated verification algorithm is used to conduct signature verification and storage operations. When compared to traditional IoT data transmission and centralized storage solutions, the newly developed approach allows for the integration of IoT and blockchain, taking full advantage of the decentralization, high dependability, and low price of blockchain to securely convert and store users' image records [33].

The solution's stability is shown via a safety study, which also demonstrates how it can protect user picture information while it is being sent and stored. When considering the smart grid application concept, the writers propose a blockchain design based on the usage of sidechains in attempt to create the network accessible and flexible, which they believe would be beneficial. The writers chose three blockchains to guarantee privacy, safety, and confidence in the entire system, and they did so with the help of the community. Moreover, in order to make the suggested solution universally applicable, they implemented the open smart grid protocol as well as smart agreements. The shown findings demonstrate how confidentiality is ensured via the suggested architecture, demonstrating that it is practical for implementation in real-world systems and applications [34]. A blockchain-based trust monitoring system with a lightweight consensus mechanism is proposed by the researchers [35]. An example of a real item connected to IoT is a car. Other examples of actual items connected to the IoT involve houses, manufacturing machines, and medical detection gadgets [36]. This very connection would make use of electrical components such as detectors and controllers in order to accomplish its goals: to share and latest data, and therefore to negotiate with the entire system in order to accomplish the best possible functionality.

4.5.5 Market Size

In their prediction [37], Gartner estimates that now the IoT will have 26 billion devices deployed by 2020, after which point providers of IoT goods and offerings would earn additional revenue surpassing $300 billion, primarily in the form of services. Naturally, companies across all industries are excited about the potential of the IoT. The Internet of Everything, as described by Cisco Systems, is a combination of IoT and more conventional IT gadgets (PCs, laptops, cell phones, and so on). The Industrial Internet, for example, is a term used by General Electric to describe the worldwide network that connects individuals, knowledge, and equipment. Furthermore, according to CEO Jeff Immelt [38], the Industrial Internet holds the ability to contribute $10 trillion to $15 trillion to the GDP (global domestic product) between 2012 and 2032. GE is spending $1 billion to explore Industrial Internet technology and services that are linked to the IoT.

4.5.6 Standards

The adoption of public protocols by suppliers and consumers will be critical to the success of the IoT. These protocols will enhance gadget supervision and administration, big data content collecting and analysis, and entire network connectivity. The IEEE Standards Association is currently working on a number of important IoT standards projects (IEEE-SA). Additionally, vendor-led organizations such as the All Seen Alliance are pushing free software initiatives to expand the IoT further than the linked home. Consumer electronics firms, household appliance manufacturers, car manufacturers, cloud service providers, retailers, software programmers, and other businesses are among the more than 100 members.

4.5.7 Security

Considerations as per Gartner [39], the growing digitalization and automation of the multitudes of gadgets placed throughout various regions of contemporary urban settings will provide new safety concerns to numerous businesses because of the growing digitization and automation of this equipment

4.5.8 People and Process Considerations

Major privacy threats will continue to exist as a consequence of the large amount of content generated as a response of the installation of a large number of equipment, which will significantly raise security difficulty. That in response will have an effect on dependability needs, which are anticipated to rise as a result, according to Gartner [40], placing actual corporate operations, as well as individual wellbeing, in jeopardy.

4.5.9 Consumer Privacy Considerations

A huge quantity of material will be generated, as is currently the case with intelligent monitoring gear and more digitalized cars, giving knowledge on individuals' personalized usage of gadgets that, if not properly protected, may lead to intrusions of security. This is especially difficult since, according to Gartner, the knowledge produced by IoT equipment is critical to providing improved services and managing such gadgets in the future.

4.5.10 Data Management Considerations

Consumer-driven individual information and big data are the two kinds of data that will be stored as a result of the IoT's effect on storage (enterprise-driven). According to Gartner [41], large amounts of data will be produced as applications and gadgets grow to educate about their clients.

4.5.11 Storage Management Considerations

These are some of the factors adding to the growing need for additional production ability is the effect of the IoT on collection architecture. This is a problem that will have to be tackled as more and more content is generated. According to Gartner, the emphasis now must be on collection capacity as well as whether or not the company can collect and utilize IoT knowledge in a cost-effective way.

4.5.12 Server Investment Considerations

According to Gartner, the IoT will have a major effect on the database industry, with increasing expenditure concentrated on key vertical sectors and organizations linked to those companies where IoT may be lucrative or provide major worth to the bottom line.

4.5.13 Bandwidth Considerations

Human engagements with algorithms produce modest capacity needs, which are met by the present data warehouse wide area network (WAN) connections. The IoT pledges to significantly alter these patterns by sending vast quantities of tiny signal device information to the data warehouse for analysis, resulting in a huge increase in incoming traffic.

4.6 IoT Enabling Technologies

1. Radio Frequency Identification (RFID) was the most important innovation of the 2000s. Afterwards, however, Near Field Communication (NFC) emerged as the dominant technology. During the 2010s, it gained more popular in cell phone devices, where it was used to scan NFC tags or to get permission to various devices such as security gates, merchant transaction systems, and so on.

2. The advancement of optical devices such as Li-Fi (Light Fidelity) or Cisco's 40 GBPS bidirectional Internet (BiDi) [42] may be beneficial for the IoT.

3. Low-cost labeling may be accomplished via the use of QR tags. Image computing is used to decipher QR codes by smartphone cameras.

4. Because of its features such as reduced power usage and data percentage, as well as its cheap and high signal bandwidth, ZigBee communication innovation, which is based on the IEEE 802.15.4 2.4-GHz radio communication protocol, is poised to become an amusing IoT enabler technology.

5. IoT interaction may be enabled via the use of Ethernet over twisted pair or fiber-optic network connections.

6. It is an upgraded version of the original LTE (Long Term Evolution) technology, with more coverage, better speed, and lower latency than the original LTE technology. It is intended for use in telecommunications, but it is also capable of being used in vehicle-to-vehicle (V2V) transmission.

7. Wi-Fi-Direct, a wireless network for peer-to-peer connectivity that does not need the use of a gateway, enables the development of IoT applications that have the throughput of Wi-Fi while experiencing reduced latency.

8. Z-Wave is the most widely utilized connectivity standard in the context of smart home devices. It makes use of a radio method operating in the 900 MHz frequency. Targeting primarily the home management and automation sector, Z-Wave technology offers a simple and dependable way of wirelessly controlling lights, HVAC (heating, cooling, and air conditioning), safety equipment, and other devices. In order to operate a Z-Wave automation network from anywhere in the world through the Internet, a Z-Wave gateway or central control equipment must be installed, which serves as both the Z-Wave hub operator and a portal to the outside world [43].

9. Thread is a communication interface for the IoT that is founded on IPv6. Thread makes use of 6LoWPAN (IPv6 over Low Power Wireless Personal Area Networks), which in turn makes use of the IEEE 802.15.4 wireless standard, which is similar to ZigBee in that it utilizes mesh connectivity. Thread, on the other hand, is IP addressable, has cloud access, and uses AES (Advanced Encryption Standard) encryption. It can now handle up to 250 gadgets in a local connectivity mesh [44], which is a significant increase over previous versions.

4.6.1 Authentication and Authorization in IoT

Several researchers have identified this as a crucial moment in the development of system protection. Interlinked equipment inside the IoT ecosystem is needed for the establishment of encrypted interaction with the assistance of appropriate verification processes. A central power in the IoT system, depending on the public key infrastructure (PKI), is usually responsible for maintaining the authenticity and authorization procedures of the linked stations and equipment [45]. As a result, the procedure substantially raises the burden of the competence unit and creates severe delays because of the high quantity of demands [46]. Several different validation approaches have been suggested in order to achieve this goal. The identification and confidentiality technique described in paper [47] is based on IP-Sec and Transport Layer Security, and it is a combination of the two (TLS). However, owing to the significant requirement on computing resources, such a method is not appropriate for resource-constrained linked IoT equipment that is linked. In the IoT system, development is developing an access control approach that is founded on a decentralized blockchain design. Research offers a system for verifying IoT that includes layers, intersects, and self-organizing blockchain structures (BCS). The storage efficiency, reaction speed, and verification of the model are all included in the evaluation of security capability [48].

4.7 Research Challenges and Future Trends

The 5G network offers characteristics that can meet the needs of the upcoming years IoT; nevertheless, it as well brings up a latest series of fascinating scientific problems in areas such as the design of 5G-IoT networks, trustworthy connectivity among equipment, privacy concerns, and other related topics. Incorporating a variety of innovations, 5G-IoT is having a major effect on the IoT that are being developed. During this part, we'll go over some of the possible study difficulties and coming year's developments in 5G-IoT technology.

4.7.1 Technical Challenges

Despite the fact that considerable scientific activities have been devoted to 5G-IoT, there are indeed technological hurdles to overcome:

1. 5G-IoT infrastructure is a significant issue, as discussed previously. A variety of structures have been suggested, each with a variety of benefits; nevertheless, the creation of the framework still poses many difficulties, including the following:

2. Scalability and connection control, because of the large quantity of IoT gadgets, scaling and connection control are significant concerns in the 5G-IoT. Handling the current details of a large amount of IoT gadgets is another problem that must be taken into consideration [49,50].

3. It is very difficult to provide smooth connectivity across heterogeneous systems due to their information sharing and heterogeneity. It is anticipated that a large quantity of IoT gadgets would be linked via communication gear in order to interact, distribute, and gather critical data with several other intelligent systems or programs [51,52].

4. Issues about safety guarantee and confidentiality, as well as safety and cyber-attacks, as well as increasing confidentiality fears.

5. The efficacy of radio software defined networks (SDNs) for 5G information communication is indeed a question mark for researchers. Despite the fact that SDN has increased scalability, there are indeed technological holes that require to be addressed.

6. The adaptable SD-CN presents a problem in terms of network scalability since it is designed to offer the main system with great levels of agility.

7. For the majority of SDN, the segregation of the controlling and information planes is problematic.

Many current IoT programs are comprised of overlay installations of IoT equipment systems, in which both the gadgets and the apps are incapable to communicate with and exchange data with one another. Meanwhile, the possibility and efficacy with which information can be collected and distributed in the natural universe continue to be a challenge. A multilayer and multidimensional solution supply system for the IoT is suggested, which solves both of the difficult problems discussed earlier. Many problems, such as the installation of dense heterogeneous systems in the IoT, the use of various accessibility methods in 5G and above 5G networks, the simultaneous delivery of full-duplex data, and so on, remain unsolved [53].

4.7.2 Security Assurance and Privacy Concerns

In order to handle complicated programs such as intelligent cities, intelligent networks, and other IoT services, important new safety abilities will be required at the gadget and network layers in the future 5G-IoT system. The safety of the varied 5G-IoT network is very difficult to manage. Not only should the developer address program infiltration from a distance, but he or she should also analyze regional invasion at the equipment [54]. However, safety guarantee must take into account the need to prevent poor safety connections.

1. Identification
2. Verification
3. Guarantee
4. Key service
5. Cryptography method
6. Transportation
7. Warehousing
8. Reverse support
9. Security
10. Authenticity

4.7.3 Standardization Issues

Inside this 5G-IoT, a great number of IoT services will offered. Because of the standards of 5G-IoT, the implementation and creation of apps will be simpler. Currently, there is indeed a shortage of uniformity and standard including both IoT platforms and services, which is related to the varied structure of networks and equipment in the 5G-IoT. There are yet numerous obstacles and challenges to overcome in order to put these ideas into action [55]. The obstacles to 5G-supported IoT standards may divided into four classifications:

1. Among the IoT equipment, or platforms, are the shape and layout of IoT goods, big data analysis applications, and so on.
2. In addition to communication networks and standards that link IoT gadgets, communication is also defined as follows.
3. Business strategies that are anticipated to meet the needs of ecommerce, vertical, horizontal, and consumption industries are being developed.

Deadly apps, which contain administrative capabilities, information collecting and analytical features, are also available.

4.7.4 Security Assurance and Privacy Concerns

1. Transmission security across 5G systems in the existence of eavesdroppers is a difficult study area to crack in the coming years. Despite formerly seen as a greater issue to be addressed via crypto techniques, gadget safety is evolving as a potential means of protection in the pursuit of 5G-IoT implementation.

2. In addition, the safety design will emphasize new confidence forms and identification administration, service-oriented safety, safety evaluation, low movement safety, and customer confidentiality protections from the physical level to the new technologies, among other aspects.

3. Privacy protection that is power friendly, since the IoT will include millions of resource-constrained equipment that will be unable to utilize computational safety systems; lightweight safety alternatives for resource-constrained endpoints will be a major study topic.

IIoT networks have needs that are distinct from those of cryptocurrency. As a result, the combination of blockchain technology with IIoT is indeed a difficult job to do. First of all and foremost, IoT equipment is asset limited. Secondly, big IIoT systems include thousands of instruments, which necessitate the use of a scalability management platform that is adequate for the channel's size. In order to solve these issues, this article offers a blockchain-based structure for the IIoT system that is lightweight, extremely expandable, and completely secure. The following are the most significant achievements made by the suggested structure.

This research demonstrates the enormous opportunity of combining blockchain's with the IoT in a way that is appropriate for intelligent industrial settings. A blockchain service level, which consists of two components, is introduced in order to guarantee excellent efficiency while also reducing the computational difficulty of blockchain transactions. During the first instance, ARM Cortex-M processors are used to create lightweight hubs that execute asymmetric cryptography in actual environments. As a second step, the primary blockchain system has implemented the evidence of authenticity (PoAh), which is a highly flexible and IoT-friendly agreement method. Customer and equipment identification, detector and controller file collection, and customer support activities are all performed in a reliable way using the suggested design. The suggested platform's effectiveness is measured using a variety of measures, including the number of consensus methods utilized, resource consumption, efficiency, and application processing duration. The suggested methodology is then applied in a fruit handling facility as an industrial trial, and the results are reported.

4.8 Conclusion

The distributed database tracks a growing list of records called blocks to prevent tampering and damage or modification of the specific blocks. When you mix blockchain with IoT, you get two far larger transactions for a number of reasons than the internet.

Transportation of consignment, monitoring and compliance with components, record operating maintenance data are some of the areas where IoT and blockchain may be felt. Some of the advantages that may be derived from security, flexibility and data management, etc. Blockchain and the IoT idea of technical characteristics, as we know, IoT work in this continuing age is no longer just a concept. It's the demand of time in everyday life. In daily life, smartphone is one of the most famous IoT activities. This allows to transmit many data kinds across various applications and hardware. IoT's not for smart houses alone. It covers all aspects from business and trade to agriculture, public safety, and health. It also includes technical ideas describing what it is, explaining what blockchain enables, and its advantages and possible problems for future thinking. It also provides an example of an application built on blockchain and talks about regulatory and governance issues.

References

[1] Retrieved form https://us.norton.com/internetsecurity-iot-what-is-the-internet-of-things.html#:~:text=How%20do%20IoT%20devices%20work%3F&text=IoT%20devices%20contain%20sensors%20and,vulnerable%20to%20malware%20and%20hacking

[2] Retrieved form https://www.pcmag.com/picks/the-best-indoor-home-security-cameras

[3] Retrieved form https://www.notebookcheck.net/Google-Glass-Enterprise-Edition-details-leak-out.146184.0.html

[4] Retrieved form https://www.meddeviceonline.com/doc/gas-sensing-smart-capsule-monitors-stomach-health-0001

[5] Retrieved form https://www.avnetworkshop.com/IoT-product-lifecycle-management.html

[6] Retrieved form https://www.geeksforgeeks.org/advantages-and-disadvantages-of-iot/

[7] Retrieved form https://vilmate.com/blog/the-internet-of-things-future-is-coming-5-iot-trends-for-2018/

[8] Guillemin, P. F. P. (2009). The Industrial Internet of Things Volume G1: Reference Architecture. [Online]. Available: http://www.internet-ofthings-research.eu/pdf/IoTClusterStrategicResearchAgenda2009.pdf

[9] Hatton, M., "The global M2M market in 2013," London, U.K., Machina Research, White Paper, Jan. 2013.

[10] Zhou, Y., F. R. Yu, J. Chen and Y. Kuo, "Cyber-Physical-Social Systems: A State-of-the-Art Survey, Challenges and Opportunities." *IEEE Communications Surveys and Tutorials* 22 (2020) (1): 389–425.

[11] Nawir, M., A. Amir, N. Yaakob and O. B. Lynn, "Internet of Things (IoT): Taxonomy of Security Attacks," In Proceeding 3rd International Conference Electronic Design (ICED), 2016, pp. 321–326.

[12] Stiawan, D., M. Idris, R. F. Malik, S. Nurmaini, N. Alsharif and R. Budiarto, "Investigating Brute Force Attack Patterns in IoT Network." *Journal of Electrical and Computer Engineering* 2019 (2019), Art. no. 4568368. [Online]. Available: https://www.hindawi.com/journals/jece/2019/4568368/

[13] Joshi, N. (2019). 8 Types of Security Threats to IoT. [Online]. Available: https://www.allerin.com/blog/8-types-of-security-threats-toiot

[14] 5G Security Recommendations Package# 2: Network Slicing, NGMN Alliance, Frankfurt, Germany, 2016

[15] Sattar, D. and A. Matrawy, "Towards Secure Slicing: Using Slice Isolation to Mitigate DDoS Attacks on 5G Core Network Slices," In Proceeding IEEE Conference Communication Network Security (CNS), 2019, pp. 82–90.

[16] Bhardwaj, K., J. C. Miranda and A. Gavrilovska, "Towards IoT-DDOS prevention using edge computing," In Proceeding {USENIX} Workshop Hot Topics Edge Computing (HotEdge), 2018. [Online]. Available: https://www.usenix.org/system/files/conference/hotedge18/hotedge18-papers-bhardwaj.pdf

[17] Zakaria, O., J. Britt and H. Forood, "Internet of Things (IoT) automotive device, system, and method," U.S. Patent 9 717 012, Jul. 25, 2017

[18] Industry 4.0: The Fourth Industrial Revolution—Guide to Industrie 4.0. Accessed: Jul. 14, 2020. [Online]. Available: https://www.i-scoop.eu/industry-4-0/#:~:text=Industry%204:0%0is%20the%20digital;of%20the%20industrial%20value%20chain

[19] Qiu, T., J. Chi, X. Zhou, Z. Ning, M. Atiquzzaman and D. O. Wu, "Edge Computing in Industrial Internet of Things: Architecture, Advances and Challenges." *IEEE Communications Surveys and Tutorials* 22 (2020) (4): 2462–2488.

[20] Gurtov, A., M. Liyanage and D. Korzun, "Secure Communication and Data Processing Challenges in the Industrial Internet." *Baltic Journal of Modern Computing* 4 (2016) (4): 1058–1073.

[21] Qadri, Y. A., A. Nauman, Y. B. Zikria, A. V. Vasilakos and S. W. Kim, "The Future of Healthcare Internet of Things: A Survey of Emerging Technologies." *IEEE Communications Surveys and Tutorials* 22 (2020) (2): 1121–1167.

[22] Catarinucci, L. et al., "An IoT-Aware Architecture for Smart Healthcare Systems." *IEEE Internet Things Journal* 2 (2015) (6): 515–526.

[23] Ranaweera, P. S., M. Liyanage and A. D. Jurcut, "Novel MEC Based Approaches for Smart Hospitals to Combat COVID-19 Pandemic." *IEEE Consumer Electronics Magazine* 10 (2021) (2): 80–91.

[24] Hossain, M. S. and G. Muhammad, "Cloud-Assisted Industrial Internet of Things (IIoT)—Enabled Framework for Health Monitoring." *Computer Networks* 101 (2016): 192–202.

[25] Chang, Y., X. Dong and W. Sun, "Influence of Characteristics of the Internet of Things on Consumer Purchase Intention." *Social Behavior and Personality: An International Journal* 42 (2014) (2): 321–330.

[26] Theodoridis, E., G. Mylonas and I. Chatzigiannakis, "Developing an IoT Smart City Framework," In Proceeding IISA, 2013, pp. 1–6.

[27] Smetanin, S., A. Ometov, M. Komarov, P. Masek and Y. Koucheryavy, "Blockchain Evaluation Approaches: State-of-the-Art and Future Perspective." *Sensors* 20(2020): 3358.

[28] Li, H., C. Lan, X. Fu, C. Wang, F. Li and H. Guo, "A Secure and Lightweight Fine-Grained Data Sharing Scheme for Mobile Cloud Computing." *Sensors* 20 (2020): 4720.

[29] Khan, P. W., Y.-C. Byun, N. Park, "IoT-Blockchain Enabled Optimized Provenance System for Food Industry 4.0 Using Advanced Deep Learning." *Sensors* 20 (2020): 2990.

[30] Lupascu, C., A. Lupascu and I. Bica, "DLT Based Authentication Framework for Industrial IoT Devices." *Sensors* 20(2020): 2621.

[31] Li, Y., K. Ouyang, N. Li, R. Rahmani, H. Yang and Y. Pei, "A Blockchain-Assisted Intelligent Transportation System Promoting Data Services with Privacy Protection." *Sensors* 20 (2020): 2483.

[32] Jamil, F., S. Ahmad, N. Iqbal and D.-H. Kim, "Towards a Remote Monitoring of Patient Vital Signs Based on IoT-Based Blockchain Integrity Management Platforms in Smart Hospitals." *Sensors* 20 (2020): 2195.

[33] Li, Y., Y. Tu, J. Lu and Y. Wang, "A Security Transmission and Storage Solution about Sensing Image for Blockchain in the Internet of Things." *Sensors* 20 (2020): 916.

[34] Sestrem Ochôa, I., L. Augusto Silva, G. de Mello, N. M. Garcia, J. F. de Paz Santana, V. R. Quietinho Leithardt, "A Cost Analysis of Implementing a Blockchain Architecture in a Smart Grid Scenario Using Sidechains." *Sensors* 20 (2020): 843.

[35] Lwin, M. T., J. Yim and Y.-B. Ko, "Blockchain-Based Lightweight Trust Management in Mobile Ad-Hoc Networks." *Sensors* 20 (2020): 698.

[36] Brown, Eric (2016). "21 Open Source Projects for IoT," Linux.com. https://www.linux.com/news/21-open-source-projects-iot/

[37] "Gartner Says 6.4 Billion Connected "Things" Will Be in Use in 2016, Up 30 Percent From 2015," *Gartner*. 10 November 2015.

[38] "Industrial Internet: Pushing the Boundaries of Minds and Machines," – Peter C. Evans and Marco Annunziata.

[39] "Gartner Says the Internet of Things Will Transform the Data Center," – Gartner. 19 March, 2014.

[40] "Gartner Says the Internet of Things Will Transform the Data Center," – *Gartner*. 19 March, 2014.

[41] "Gartner Identifies the Top 10 Strategic Technology Trends for 2015," – Gartner. 8 October, 2014.

[42] "Top Five Things to Know about Cisco BiDi Optical Technology White Paper"- Cisco. 5 September 2014. "40-Gbps bidirectional (BiDi) optical technology".

[43] "Smarten up your dumb house with Z-Wave automation," Digital Trends.

[44] "Introducing Thread," SI Labs.

[45] Cooper, D., S. Santesson, S. Farrell, S. Boeyen, R. Housley and W.T. Polk, "Internet X.509 Public Key Infrastructure Certificate and Certificate Revocation List (CRL) Profile." *RFC* 5280 (2008): 1–151.

[46] Oriwoh, E. and M. Conrad, "'Things' in the Internet of Things: Towards a Definition." *International Journal of Internet of Things* 4 (2015): 1–5.

[47] Ukil, A., S. Bandyopadhyay and A. Pal, "Iot-privacy: To be private or not to be private," In Proceedings of the 2014 IEEE Conference on Computer Communications Workshops (INFOCOM WKSHPS 2014), Toronto, ON, Canada, 27 April–2 May 2014, pp. 123–124.

[48] Qu, C., M. Tao, J. Zhang, X. Hong and R. Yuan, "Blockchain Based Credibility Verification Method for IoT Entities." *Security and Communication Networks* 2018 (2018). Article ID7817614. https://doi.org/10.1155/2018/7817614

[49] Mach, P., Z. Becvar and T. Vanek, "In-Band Device-to- Device Communication in OFDMA Cellular Networks: A Survey and Challenges." *IEEE Communications Surveys & Tutorials* 17 (4): 2015: 18851922.

[50] Ndiaye, M., G. P. Hancke and A. M. Abu-Mahfouz, "Software Defined Networking for Improved Wireless Sensor Network Management: A Survey." *Sensors* 17 (2017) (5): 1–32.

[51] Ishaq, I. et al., "IETF Standardization in the Field of the Internet of Things (IoT): A Survey." *Journal Sensor Actuator Network* 2 (2013) (2): 235–287.

[52] Elkhodr, M., S. Shahrestani and H. Cheung, "The Internet of Things: New Interoperability, Management and Security Challenges," arXiv preprint arXiv:1604.04824, 2016.

[53] Project CONTENT FP, 2012–2015, [Available on line 15 Jan 2018], http://cordis.europa.eu/fp7/ict/future-networks/

[54] Zhao, Shuai, Le Yu and Bo Cheng, "An Event-Driven Service Provisioning Mechanism for IoT (Internet of Things) System Interaction." *IEEE Access* 4 (2016) (2): 5038–5051.

[55] Girson, Andrew, "IoT Has a Security Problem Will 5G Solve It?", [Available on line 15 Jan 2018], https://www.wirelessweek.com/article/2017/03/iot-has-securityproblem-will-5g-solve-it

5

Blockchain Implementation Challenges for IoT

Trishit Banerjee

Netaji Subhash Engineering College, Techno City
Garia Kolkata, India

CONTENTS

5.1 Introduction

Blockchain is a widely known technology that is currently used in different business sectors. It is considered a disruptive technological innovation that enables to revolutionize the trading and interaction process of society. This reputation is in specific attributable to its characteristics, which enables mutually mistrusting entities in conducting financial transactions and interacting without depending on a reliable third party. Blockchain technology and distributed ledgers are gathering immense attention and initiating numerous projects in various industries. However, the financial industry can be considered the key user of the blockchain system [1]. This particular technology was initially created for supporting the renowned cryptocurrency Bitcoin in 2009. Since then, the application of this technology has been gradually increasing in different aspects of the business. Blockchain usually refers to a chain of blocks that save all committed transactions with the help of a public ledger. The length of the chain constantly increases while new blocks are attached to it.

Moreover, blockchain functions in a decentralized environment that includes different core technologies: "cryptographic hash," distributed consensus algorithms, and "digital signatures." All the financial transactions happen in a decentralized way that eradicates the necessity for intermediaries to verify and validate the transactions. The technology has major characteristics: "transparency," "decentralization," "audibility, and "immutability" [2]. Apart from Bitcoin, the technology can be applied in other financial services, such as remittance, digital assets, and online payments. It can be applied in supply chain management, healthcare and hospitality management, digital media transfer, and many more. However, numerous kinds of challenges can emerge during the implementation of this technology. This paper will strongly focus on those challenges limiting the use of blockchain along with its structure and different application areas.

5.2 Discussion of Challenges Regarding the Implementation of Blockchain

Despite the wide range of blockchain applications in multiple business areas, several crucial challenges remain. These challenges limit the use of blockchain as well as prevent it from being applied in digital trades. Some of the challenges are described as follows.

5.2.1 Scalability

It is a vital challenge related to the implementation of blockchain as it limits the functionality of the technology. The size of the blockchain has become significantly huge as the usage volume of the technology is increasing, and the number of occurred transactions is rising daily—blockchain processes and stores thousands of transactions in nodes to obtain validation [3]. Primarily, the present transaction source requires to be confirmed before the transaction itself. Moreover, the time break and block size limitation necessary for generating new blocks maintain a significant role in not fulfilling requirements regarding the instantaneous processing of many transactions in a real-time environment. Apart from that, the size of each block can also be an issue that is responsible for delaying the transaction.

5.2.2 Integrated Cost

High costs related to the implementation of blockchain can also be another critical challenge for most organizations. Replacing the old and existing system with a blockchain-centric new system requires huge financial resources and time as it involves many upgraded technologies. In addition, discovering the most appropriate blockchain system is not easy as most of them are not entirely developed and tested [4]. Additionally, even the most reasonable platforms also require a huge cost to be implemented along with energy cost. There is no specific cost for a blockchain-based solution as its price depends on the organization's services that plan to implement it.

5.2.3 Privacy and Security Challenges

Privacy and security measures are some of the biggest concerns regarding the implementation of blockchain. Leakage of privacy related to the transaction is a major vulnerability of the technology. The balances and the other credentials of public keys can be visible to those connected to that network. Due to that reason, there is a high chance of data breach, and as the blockchain is mostly used in a financial transaction, the risk factors can enhance to a huge extent [5]. This behavior makes the technology less applicable in most cases. However, this problem can be resolved by obtaining anonymity in the blockchain. It can be categorized into a mixing solution and an anonymous solution.

5.2.4 Regulation Issues

The lack of rules and regulations or improper guidelines and regulations is another major constraint in blockchain implementation as those are different in different countries. Keeping up with the technology advancement has always been a struggling factor for regulations and guidelines. However, technology like blockchain tends to evade regulations completely for tackling incompetence associated with the payment networks [6]. This technological approach's major challenge is it reduces oversight, which was further one of its unique motivations [7]. Major governments are yet unclear regarding their perception of blockchain from a legal position. The complexity of the technology is responsible for this situation. For that reason, potential users may face legal issues in the future, which can limit them from using this technology. The concerned authority needs to focus on this issue and formulate new strategies and policies.

5.2.5 Energy Consumption

It is a significant issue that restricts the usage of blockchain in specific sectors. For validating transactions, the technology mainly functions on the proof-of-work (PoW) mechanism. Moreover, this particular mechanism needs complex mathematical calculations for validating transactions [8]. In this way, these computations need sufficient energy to deliver power to computers operating this task. Therefore, the energy consumption related to the blockchain implementation is very high, which also enhances its operational cost. Besides, this higher level of energy consumption for functioning is a constraint to the majority of the businesses that are now considering other energy sources or other sustainable ways to operate.

5.2.6 Lack of Standardization

A lack of standardization or limited interoperability among many networks is another crucial challenge that confines the adaptation of blockchain. More than 6,500 projects utilize a particular variety of blockchain platforms, mainly stand-alone, with solutions comprising various protocols, consensus mechanisms, coding languages, and privacy measures. The space of blockchain is in a "state of disarray" as there are so many different networks available [9]. It occurs because of a lack of universal standards that would enable various networks to communicate. The absence of such equivalence among blockchain standards further reduces consistency from fundamental procedures such as security, making mass implementation a nearly impossible task. There are different categories of blockchain networks, such as private, consortium, and public. Based on application requirements, each of the networks has its disadvantages and advantages. Organizations cannot rely on one specific blockchain solution as its protocols and standards vary within different industries, making it challenging to implement.

5.2.7 Selfish Mining

Selfish mining is one of the most common challenges concerning blockchain implementation. A block is susceptible to cheating against a minimal utilization of hashing power. In this manner, the miners keep the mined block without creating any broadcast on the network. However, the miners create a private branch and broadcast after fulfilling their particular requirements [10]. For this reason, resources and time are wasted by legitimate miners, whereas selfish miners can mine the private chain. The vulnerability of blockchain against coding is the primary reason for the selfish phishing attacks.

5.2.8 Lack of Understanding and Awareness

There is less knowledge available regarding the implementation and utilization of blockchain as it is still an emerging technology. The majority of people do not have consciousness about this new technology. They do not have enough skills and expertise to handle such complex technology, making it more avoidable in most organizations. The blockchain council has reported that the demand for blockchain engineers elevated to ver 500% in 2019 compared to 2020, with basic salaries for blockchain developers enhancing alongside [11]. Therefore, the technical maturity of blockchain is still very low as this particular technology is still in the developing stage. Many technology functions are yet to be discovered, making them applicable to areas other than cryptocurrency and business context.

Along with that, blockchain is easily vulnerable to the issues such as system failure, capacity, non-predicted bugs. Most significantly, the technology is going to be used by technically unsophisticated users. Therefore, less understanding and awareness is a serious issue regarding implementing this technology.

5.3 Literature Review

5.3.1 Significance of Blockchain in Business

Gates [12] stated that the most important benefit of blockchain's distributed ledger is low operational cost. The cost can be significantly lowered by eliminating intermediaries or the system of transaction reconciliation and record keeping. Moreover, by eradicating the data gatekeeper or middleman, this particular technology enables organizations to trace products and transactions conveniently and quickly back to their roots. For instance, renowned American global retailer Walmart has decreased its overall time from 7 days to 2.2 seconds required for tracing food from stores to farms. Therefore, the technology has massively helped the retail organization to reduce the consumption of time.

Additionally, the overall cost can be hugely lowered by lowering the time consumption as well. According to Francisco and Swanson [13], blockchain can provide greater transparency to businesses, making it more reliable and authentic. Each company has to maintain a separate database if they do not use blockchain. This technology involves a distributed ledger that keeps track of data and transactions identically in several locations. All participants joined in the network with permission access can observe the data simultaneously, ensuring complete transparency.

Furthermore, all transactions recorded by blockchain are date and timestamped. Due to this, a user can be able to check the overall history of a particular transaction. It virtually eradicates any chance of fraud.

However, Casado-Vara et al. [14] argued in a study that this blockchain could also be used in business for its improved speed and efficiency. It can be noted that traditional transaction processes are highly time-consuming, have a chance of error, and frequently needs an intermediary for successful completion. On the contrary, blockchain can complete the transactions more efficiently and faster as it can process a large number of calculations simultaneously. Transaction details can be stored on the blockchain along with documentation which eradicates the requirement for exchanging paper. The technology also eliminates the need to handle numerous ledgers, making the settlement and clearing much faster. Xu, Chen, and Kou [15] opined that blockchain could also be significant for business as it allows smart trading. With the help of this technology, organizations can create smart contracts that are extensively utilized for implementing business collaboration in general and inter-organizational business operations in specific. By using blockchain, organizations can automate transactions based on smart contracts where manual confirmation is not needed. For example, companies can automatically file taxes by using smart contracts.

5.3.2 Concept of Consensus Algorithm

Pahlajani, Kshirsagar, and Pachghare [16] narrated that the consensus algorithm or layer can be considered the critical aspect in the blockchain architecture. This particular algorithm configures protocols for deciding the process of adding a new block in the blockchain. Moreover, the consensus algorithm resolves the issue of trust in the technology. However, it can also be referred to as a specific process that allows all the blockchain network participants to reach a shared agreement regarding the current state of the distributed ledger. Therefore, a consensus algorithm aspires to find a joint agreement that is acceptable to the entire network. In this manner, the algorithm helps to acquire high reliability in the network and ensures the increase of trust among unknown participants in the distributed environment.

Additionally, the consensus protocol further ensures that each added block in the blockchain is the sole version of the truth accepted by all the technology nodes. In addition, Nguyen and Kim [17] also stated in a study that consensus algorithms could be categorized into two particular classes. "Voting-based consensus" is considered the first class that helps blockchain network nodes broadcast their outcomes regarding transaction or mining a new block before adding the block in the blockchain. Moreover, the second class is "proof-based consensus" that allows the nodes joining in the network to solve a mathematical puzzle for showcasing they are more effective than the other nodes for doing the mining work.

As opined by Mingxiao et al. [18], different types of consensus algorithms are mainly used in blockchain. PoW is a popular algorithm that helps in selecting a miner regarding the next block creation. Bitcoin uses this consensus algorithm. The central notion behind this algorithm is to resolve complex mathematical riddles and provide a solution quickly. This puzzle needs high calculation power, and for that, the node that resolves the dilemma can mine the next block at once. Besides, practical Byzantine fault tolerance (PBFT) is another crucial algorithm of a distributed network that can reach consensus.

In contrast, some other network nodes become unable to respond or provide inaccurate data. The key objective of this algorithm is to provide safety against system malfunction by gathering data from both the correct and defective nodes to lower the impact of the faulty nodes. It is separated from "Byzantine Generals' Problem." Alongside, proof of stake (PoS) is another consensus algorithm that has been introduced in the first Bitcoin project but is not commonly utilized for its robustness and other causes. In this algorithm, the digital currency is based on the coin age's concept. Coin age refers to a value that is multiplied by the time limit after its generation. A node can get into the blockchain network more accurately if it holds the coin for longer.

5.3.3 Types of Blockchain

According to Niranjanamurthy, Nithya, and Jagannatha [19], there are mainly three types of blockchain that are generally used. These are "public blockchain," "private blockchain," and "consortium blockchain." It can be stated that each blockchain has its characteristics that make them useful in the business world. Public blockchains are open to the public, which allows anybody to become the network user and take part in the core operations. Moreover, any individual who can further participate in the consensus process can check and verify the transactions related to it. Sometimes, these blockchains are considered "permissionless" blockchains as there is no permission required for interacting with the protocol. However, being open to all does not make it less secure than the other blockchains. The transaction data is viewable in this blockchain, such as date, amount, and wallet number.

On the other hand, the private blockchains are opposite of the public blockchains as they are more restricted, and only certain nodes are allowed to participate in the consensus. In this management, a central authority, which can be a business or enterprise, takes control of creating and validating transactions and decides who can view the transactions. Moreover, the authority can further determine the mining rights and change or override entries on the distributed ledger. Nevertheless, in another study, Dib et al. [20] showed that the consortium blockchains could also be very effective for business operations as it exhibits a partly decentralized behavior. This blockchain can be both private and public that is regulated by specific preselected nodes. It is not fully open like a public blockchain or restricted like a private blockchain. It can be further referred to as a hybrid blockchain which shows the other two blockchain's characteristics.

5.3.4 Application of Blockchain in IoT Domains and the Specific Challenges

The primary use of the blockchain is done to make sure that there is complete security of the data. Hence, the blockchain application is made in many cases while using the Internet of Things (IoT).

There are various benefits for using the Internet of Things (IoT) in the blockchain process. According to Chamola et al. [21], the use of the blockchain is done to ensure a secure and immutable data chain that can only be added, not changed. It is also clear that there is a sufficient amount of benefit in the workplace for ensuring efficiency. Blockchain is often used in freight transportation to ensure that the movement of freight is done from one party to another one. The integration of blockchain with the Internet of Things (IoT) is essential for ensuring that temperature storage and shipping status is implemented properly. Through the introduction of the immutable blockchain, there is complete assurance of ensuring that all the parties' data are kept completely safe. Lastly, Lunardi et al. [22] have also stated that there are continuous challenges in terms of connectivity while using the Internet of Things (IoT). The IoT devices need to have a strong connection with each other to operate properly. Hence, high computing storage and networking data are considered to be partial stakeholders for implementing IoT. But on the other hand, it is also true that there is a sufficient amount of challenges regarding implementing new applications to ensure various domains.

On the other hand, Dorri et al. [23] have added that blockchain also helps ensure the implementation of tamper-resistant records of the shared transaction. For example, the IBM blockchain helps the business partners share and gain access to the IoT data present in the database without the need for control and management. The author has given the example of component tracking that helps the companies to make sure that the safety and security of both the regulatory compliance are implemented properly. There is also component tracking, and as a result of this, the implementation of the Internet of Things (IoT) is done for blockchain ledgers. Along with this, the authors Rejeb et al. [24] have also stated that the combination of the Internet of Things (IoT) and blockchain technology is essential for increasing the efficiency of the supply chain. It is also true that there is complete provenance throughout the work to ensure that there is complete sharing of information about the agencies and shippers in a cost-effective manner.

The authors Manzoor et al. [25] have also stated that the use of blockchain technology is important for ensuring the utilization of the cryptocurrency. The use of blockchain technology is stored inherently to ensure that the information regarding the transaction is made by and between the inventors. There is also the process of ensuring that the cryptocurrency blockchain stores all the information such as transaction history and ensuring that there is a large form of transcripts. Along with this, Huang et al. [26] have also pointed out the role of the blockchain on the implementation of the Bitcoin cryptocurrency. The author has stated that the main advantage of the blockchain is identified to be the high level of transparency in the workplace. There has been various form of data currency that is called blocks and also, the decentralized blockchain is also being created with the help of confirmed blocks. The blocks within the distributed ledger are considered essential for making sure that there is continuous approval of blocks with the use of cryptographic hash codes. Each of the participants in the workplace is deemed to be important for making sure that there is a peer-to-peer network that can verify the participant's behavior.

On the other hand, Singh et al. [27] stated that there are continuous challenges in using blockchain on cryptocurrency. For example, the blockchain is a very much emerging technology, and as a result, the skills needed to be developed are in short supplies. Along with this, it is also stated that there is a need for broad adaption of the clockchain technology

But on the other hand, Choi [28] states blockchain is considered a very complex process due to its high computing costs and delays. There is a high amount of power consumption and performance needed to be considered as a challenge for ensuring the advancement of technology-based applications. Along with this, it is also stated that there is a continuous form of challenge in ensuring Bitcoin in the workplace. The main challenge is that the Bitcoin network consumes a significant amount of energy more than nations such as Austria and Colombia. As a result of this, there is a continuous threat in terms of the operation of IoT. Along with this, Albayati et al. [29] have also stated that there is also a challenge in terms of blockchain processing while making use of the Internet of Things

(IoT) data. The main concern for the blockchain is that there is a continuous challenge in terms of increasing the blicks per second. For example, the complete elimination of blockchain proof of work (PoW) ensures a continuous reduction of power consumption to improve performance.

But on the other hand, there is also a challenge in terms of the Sybil attack due to the elimination of blockchain PoW. Hence, maintaining the security of the devices connected through the Internet of Things becomes more and more challenging. On the other side, Shanaev et al. [30] have also stated that there is also a challenge regarding the implementation of the data concurrency. The author has pointed out that the implementation of blockchain's throughput is very much limited due to the complex cryptographic security protocol and ensuring consensus mechanism. Along with this, it is also true that the rapid synchronization of the new blocks ensures chain structured ledger as it needs a higher level of bandwidth to improve the blockchain throughput. Hence, the increase of the blockchain's throughput is needed to meet the frequent transaction through the Internet of Things (IoT) system.

5.3.5 Role of the Blockchain as a Data Repository for AI Applications

The data depository, also known as the data library, is considered the dataset that needs to be isolated. The use of big data is considered one of the major techniques used for the collection of data. The author Hassani et al. [31] have stated that there is efficiency to make use of the blockchain when making use of the data storage. Along with this, there is also stated that there is a need to keep track of the 10,000 VHS to ensure that there is modern computing. Along with this, it is also stated to be true that blockchain-enabled big data is completely secured, and it is also clear that there is zero possibility of the data getting hacked.

Along with this, it is also stated that the blockchain-based data is very much structured and also secured. There is also the possibility of fraud detection as the financial institution can use this to check the transaction in real-time. It is stated that big data is a profitable business, and it is also expected that the annual industry for big data is expected to reach $203 billion by the year 2020. It is also likely that by the year 2030, around 20% of the entire global big data market to ensure that blockchain reaches around $100 billion in annual income.

But on the other hand, Bennett et al. [32] have also stated that there is also a challenge while using the big data regarding the network. Along with this, it is also stated that there is a continuous form of throughput for the entire peer-to-peer network with the help of decentralized storage. Even though the decentralized storage structure improves effectiveness and solves the bottleneck problem to reduce the need for third-party trust, there is a high level of data storage needed for storing the data. For example, it is stated that there is a need for approximately 730 GB of data storage for each year if there is a need for 1,000 people to exchange 2 MB of data. Along with this, it is also stated that there is a continuous form of development of the firm to ensure full benefit in the workplace.

On the other hand, Karafiloski and Mishev [33] have also stated that despite the challenges, companies need to make sure that the use of blockchain technology is done for the mineral mining process. The entire market for blockchain technology is identified to be around $700 billion, and it is also projected that there is continuous development of the market. It is expected that the use of blockchain technology is influential and is expected to grow to reach trillions. It is expected that the top 50 companies in the mining of minerals are expected to have immense profits due to the use of blockchain in a data repository.

5.4 Mathematical Foundations of Blockchain

Implementing the mathematical foundation is important for ensuring that there is complete efficiency in blockchain implementation.

One way to implement the mathematical foundation is through the help of one-way functions such as $f:\{0, 1\}^* \rightarrow \{0, 1\}^*$. This equation is considered one of the computable functions that are important for ensuring that the probability of x is found out. The value of x is found out with the random $Y = f(x)$ implementation is negligible for polynomial time.

Along with the implementation, H(x) is considered to be a hash function that is important for making sure that there is an arbitrary and finite length of the bi-string to ensure that the production of the output Y is done with fixed length (Ahir et al. [34]. pp. 113–120). It is also stated that the hash output is a random bit strong, and also there are three vital properties of the cryptographic hash function.

$$\{0, 1\}^* \rightarrow \{0, 1\}^n.$$
$$Y = H(x).$$

1. There is the use of pre-image resistance to ensure a continuous amount of output from the specific hash function. Also, there is computational resistance of finding regarding the specific amount of input is. It implies that there is stated to be difficulties within Y to ensure that it is difficult to find the value of x, such as there is a shield $Y = H(x)$ [19].

2. The second pre-image is identified as the resistance important for determining the input and ensuring that it becomes hard to find the input for any kind of hash result. This implies that even if there is x_1 and $Y = H(x_1)$, there is difficulty in terms of ensuring that finding of x_2 to ensure that $H(x_1) = H(x_2)$ [19].

3. The collision resistance implementation ensures that there are all forms of inputs and have the same amount of output. This implies that if there is $x_1! = x_2$ to ensure that there is insult within $H(x_1)! = H(x_2)$.

It is considered to be important for the firms to ensure that there is a sufficient amount of hash points.

5.5 Difference Between Blockchain and Mining Applications

Data mining is considered the process that uses exploration to ensure the application of areas and web analysis, biomedicine, and science. The use of data mining is done to ensure that there is the implementation of the anti-fraud system to analyze transactions. The role of the blockchain is also to ensure that authentication is done, but there is a complete comprising of the blockchain [35]. However, there is a need for blockchain to ensure that keeping of the transaction record is done properly. Along with this, it is also considered to be true that there is a continuous amount of information in the blockchain about how the transactions are processed. The mining process is required to have a hash of a block of transactions that cannot get forged. Along with this, it is also stated to be true that there is a sufficient amount of benefit in the workplace. Along with this, the process of protecting the integrity of the entire blockchain needs to be covered entirely.

It is also stated that there is the implementation of card transactions and improving the purchasing pattern. Along with this, it is also clear that there is a continuous form of benefit in the workplace to ensure continuous governance improvement. It is also considered true that the use of data mining is done to promote anonymity. Along with the implementation of blockchain, there is the possibility of making sure that there is additional transaction regarding the existing blockchain ledger [36]. The transaction is being distributed among all the users of the blockchain to ensure complete protection of integrity. On the other hand, the use of the blockchain is done to make sure that there is better transparency.

5.6 Discussion of Cryptographic Foundation Related to Blockchain

The blockchain is considered a growing list of records to ensure that there are different forms of algorithm such as *asymmetric-key algorithm or public key encryption* and the *hash function.* The use of the public key encryption is done to ensure a safe and secure transaction. It serves as the basis of the cryptographic hash function, blockchain wallets, and also making transactions to ensure that there is a trait for immutability. The use of the public key cryptography is done by the users to securely send the money and data ([37]. pp. 53–62). In the asymmetric key cryptography, the sender makes use of the public key to transfer the money. And also, the recipient can make access to the funds by making use of the public key that they hold. With the use of this channel, it is also stated that there is a sufficient amount of challenges in the workplace.

The use of the hash function is done for generating functionality regarding a single view for each participant. The blockchain does the uses the SHA-256 hashing algorithm to ensure the betterment of the hash function. The SHA-256 algorithm is identified to be used by the digital signature and authentication to ensure complete security. Implementing the secure hashing algorithm through the digital signature and authentication process is important for ensuring security. In the case of the SHA-256 blockchain, irrespective of the number of people, there is a continuous form of benefit in the workplace to ensure efficiency [38]. Hence, it is true that there is a continuous amount of benefit in the workplace for ensuring efficiency.

The main reason for using the SHA-256 algorithm is considered to be the use of the hashing function. Along with this, it is also stated that there is a cryptographic hash function to ensure the following benefits for blockchain:

- Every part of the data becomes unique, and as a result of this, utility goes higher.
- There is also the avalanche effect as a minor change in the hash function of the data generates very much different output.
- There is also quickness in which the output can be generated to ensure a smaller amount of time [39].
- It is also true that there is no possibility of reverse engineering in the case of blockchain hash function as the output cannot be returned to input.

The use of the hash function is done to ensure that there is a major role in ensuring that there is the continuous maintenance of blockchain [40].

5.7 Conclusion

Based on the above assessment, it is clear that the use of blockchain has an advantage and challenges when making use in the workplace. Blockchain is used in the IoT due to the improvement in the security system among the device connected with the Internet. However, due to the complex architecture, there is a challenge in implementation as this requires a significant amount of skills. It is also considered true that there is the use of the blockchain in terms of cryptocurrency and data repository.

Along with this, it is also true that there is a continuous form of benefit in terms of ensuring an efficient amount of benefit. The use of the public key cryptography and SHA-256 algorithm is also done to ensure the protection of the financial amount that is being transmitted. The main advantage of the SHA-256 algorithm is that this is the only cryptographic algorithm that hackers have not breached.

References

[1] Nofer, M., P. Gomber, O. Hinz and D. Schiereck, "Blockchain." *Business & Information Systems Engineering* 59 (2017) (3): 183–187. 10.1007/s12599-017-0467-3

[2] Monrat, A. A., O. Schelén and K. Andersson, "A Survey of Blockchain from the Perspectives of Applications, Challenges, and Opportunities." *IEEE Access* 7 (2019): 117134–117151. 10.11 09/ACCESS.2019.2936094

[3] Sharma, T. "5 Key Challenges for Blockchain Adoption In 2020," Blockchain Council. Retrieved 18 August 2021. 2020. https://www.blockchain-council.org/blockchain/5-key-challenges-for-blockchain-adoption-in-2020/

[4] Geroni, D. (2020). *"Top 10 Enterprise Blockchain Implementation Challenges,"* 101 Blockchains. Retrieved 18 August 2021. https://101blockchains.com/enterprise-blockchain-implementation-challenges/

[5] Fabiano, N., "Internet of Things and Blockchain: Legal Issues and Privacy. The Challenge for a Privacy Standard," In 2017 IEEE International Conference on Internet of Things (iThings) and IEEE Green Computing and Communications (GreenCom) and IEEE Cyber, Physical and Social Computing (CPSCom) and IEEE Smart Data (SmartData), 2017, June, pp. 727–734. IEEE. 10.1109/iThings-GreenCom-CPSCom-SmartData.2017.112

[6] Kumutha, K. and S. Jayalakshmi, "Dispelling Challenges and Applications of Blockchain Technology." *Parishodh Journal* 9 (2020) (3): 6397–6401. https://www.researchgate.net/profile/Kumutha-Krishnan-3/publication/352547798_Dispelling_Challenges_and_Applications_of_Blockchain_Technology/links/60ce8b8d299bf1cd71e147f0/Dispelling-Challenges-and-Applications-of-Blockchain-Technology.pdf

[7] Iredale, G. (2020). *"Top 10 Blockchain Adoption Challenges,"* 101 Blockchains. Retrieved 18 August 2021, from https://101blockchains.com/blockchain-adoption-challenges/.

[8] Tar, A. (2018). *Proof-of-Work, Explained*. Cointelegraph. Retrieved 18 August 2021, from https://cointelegraph.com/explained/proof-of-work-explained

[9] De Meijer, C. (2016). *Remaining challenges of blockchain adoption and possible solutions*. Finextra Research. Retrieved 18 August 2021, from https://www.finextra.com/blogposting/18496/remaining-challenges-of-blockchain-adoption-and-possible-solutions

[10] Meva, D., "Issues and Challenges with Blockchain: A Survey." *International Journal of Computer Sciences and Engineering* 6 (2018): 488–491. https://www.researchgate.net/profile/Dr-Divyakant-Meva/publication/330384987_Issues_and_Challenges_with_

Blockchain_A_Survey/links/5c3da807299bf12be3c8d2ca/Issues-and-Challenges-with-Blockchain-A-Survey.pdf

[11] Brown, M. (2021). *"5 Challenges with Blockchain Adoption and How to Avoid Them,"* SearchCIO. Retrieved 18 August 2021, from https://searchcio.techtarget.com/tip/5-challenges-with-blockchain-adoption-and-how-to-avoid-them

[12] Gates, L. (2019). *"Why Is Blockchain Important to Business?,"* Insight. Retrieved 18 August 2021, from https://www.insight.com/en_US/content-and-resources/2019/2202019-why-blockchain-is-important-to-business.html

[13] Francisco, K. and D. Swanson, "The Supply Chain Has no Clothes: Technology Adoption of Blockchain for Supply Chain Transparency." *Logistics* 2 (2018) (1): 2. 10.3390/logistics2010002

[14] Casado-Vara, R., P. Chamoso, F. De la Prieta, J. Prieto and J. M. Corchado, "Non-Linear Adaptive Closed-Loop Control System for Improved Efficiency in IoT-Blockchain Management." *Information Fusion* 49 (2019): 227–239. 10.1016/j.inffus.2018.12.007

[15] Xu, M., X. Chen and G. Kou, "A Systematic Review of Blockchain." *Financial Innovation* 5 (2019) (1): 1–14. 10.1186/s40854-019-0147-z

[16] Pahlajani, S., A. Kshirsagar and V. Pachghare, "Survey on Private Blockchain Consensus Algorithms." In 2019 1st International Conference on Innovations in Information and Communication Technology (ICIICT), 2019, April, pp. 1–6. IEEE. 10.1109/ICIICT1.2019.8741353

[17] Nguyen, G. T. and K. Kim, "A Survey about Consensus Algorithms used in Blockchain." *Journal of Information Processing Systems* 14 (2018) (1): 101–128. 10.3745/JIPS.01.0024

[18] Mingxiao, D., M. Xiaofeng, Z. Zhe, W. Xiangwei And C. Qijun, "A Review On Consensus Algorithm of Blockchain." In 2017 IEEE International Conference on Systems, Man, and Cybernetics, 2017, October, pp. 2567–2572. IEEE. 10.1109/SMC.2017.8123011

[19] Niranjanamurthy, M., B. N. Nithya and S. Jagannatha, "Analysis of Blockchain Technology: Pros, Cons and SWOT." *Cluster Computing* 22 (2019) (6): 14743–14757. 10.1007/s10586-018-2387-5

[20] Dib, O., K. L. Brousmiche, A. Durand, E. Thea and E. B. Hamida, "Consortium Blockchains: Overview, Applications and Challenges." *International Journal On Advances in Telecommunications* 11 (2018) (1&2): 51–64. https://www.researchgate.net/profile/Omar-Dib-3/publication/328887130_Consortium_Blockchains_Overview_Applications_and_Challenges/links/5be99602299bf1124fce0ab9/Consortium-Blockchains-Overview-Applications-and-Challenges.pdf

[21] Chamola, V., V. Hassija, V. Gupta and M. Guizani, "A Comprehensive Review of the COVID-19 Pandemic and the Role of IoT, Drones, AI, Blockchain, and 5G in Managing its Impact." *IEEE Access* 8(2020): 90225–90265.

[22] Lunardi, R. C., R. A. Michelin, C. V. Neu, H. C. Nunes, A. F. Zorzo and S. S. Kanhere, "Impact of Consensus on Appendable-Block Blockchain for IoT," In Proceedings of the 16th EAI International Conference on Mobile and Ubiquitous Systems: Computing, Networking and Services, 2019, November, pp. 228–237.

[23] Dorri, A., S. S. Kanhere, R. Jurdak and P. Gauravaram, "LSB: A Lightweight Scalable Blockchain for IoT Security and Anonymity." *Journal of Parallel and Distributed Computing* 134 (2019): 180–197.

[24] Rejeb, A., J. G. Keogh and H. Treiblmaier, "Leveraging the Internet of Things and Blockchain Technology in Supply Chain Management." *Future Internet* 11 (2019) (7): 161.

[25] Manzoor, A., M. Liyanage, A. Braeke, S. S. Kanhere and M. Ylianttila, "Blockchain Based Proxy Re-Encryption Scheme for Secure IoT Data Sharing," In 2019 IEEE International Conference on Blockchain and Cryptocurrency (ICBC), 2019, May, pp. 99–103. IEEE.

[26] Huang, J., L. Kong, G. Chen, M. Y. Wu, X. Liu and P. Zeng, "Towards Secure Industrial IoT: Blockchain System with Credit-Based Consensus Mechanism." *IEEE Transactions on Industrial Informatics* 15(6) (2019): 3680–3689.

[27] Singh, S., P. K. Sharma, B. Yoon, M. Shojafar, G. H. Cho and I. H. Ra, "Convergence of Blockchain and Artificial Intelligence in IoT Network for the Sustainable Smart City." *Sustainable Cities and Society* 63 (2020): 102364.

[28] Choi, T. M., "Creating all-win by Blockchain Technology in Supply Chains: Impacts of Agents' Risk Attitudes Towards Cryptocurrency." *Journal of the Operational Research Society* (2020), 1–16.

[29] Albayati, H., S. K. Kim and J. J. Rho, "Accepting Financial Transactions Using Blockchain Technology and Cryptocurrency: A Customer Perspective Approach." *Technology in Society* 62 (2020): 101320.

[30] Shanaev, S., S. Sharma, B. Ghimire and A. Shuraeva, "Taming the Blockchain beast? Regulatory Implications for the Cryptocurrency Market." *Research in International Business and Finance* 51 (2020): 101080.

[31] Hassani, H., X. Huang and E. Silva, "Banking with Blockchain-ed Big Data." *Journal of Management Analytics* 5 (2018) (4): 256–275.

[32] Bennett, R. M., M. Pickering and J. Sargent, "Transformations, Transitions, or Tall Tales? A Global Review of the Uptake and Impact of NoSQL, Blockchain, and Big Data Analytics on the Land Administration Sector." *Land Use Policy* 83 (2019): 435–448.

[33] Karafiloski, E. and A. Mishev, "Blockchain Solutions for Big Data Challenges: A Literature Review," In IEEE EUROCON 2017-17th International Conference on Smart Technologies, 2017, July, pp. 763–768. IEEE.

[34] Ahir, S., D. Telavane and R. Thomas, "The Impact of Artificial Intelligence, Blockchain, Big Data and Evolving Technologies in Coronavirus Disease-2019 (COVID-19) Curtailment," In 2020 International Conference on Smart Electronics and Communication (ICOSEC), 2020, September, pp. 113–120. IEEE.

[35] Uddin, M. A., A. Stranieri, I. Gondal and V. Balasubramanian, "A Survey on the Adoption of Blockchain in IoT: Challenges and Solutions." *Blockchain: Research and Applications* (2021): 100006.

[36] Veria Hoseini, S. and V. Niemi, *Mathematics and Data structures in Blockchain & Ethereum*, (2018). Master's Thesis under Prof. Valtteri Niemi, faculty of Mathematics and Natural Science Department of Future Technologies University of Turku August 2018.

[37] Khan, K. M., J. Arshad and M. M. Khan, "Secure Digital Voting System based on Blockchain Technology." *International Journal of Electronic Government Research (IJEGR)* 14 (2018) (1): 53–62.

[38] Cyran, M. A., "Blockchain as a Foundation for Sharing Healthcare Data."*Blockchain in Healthcare Today* 1 (2018): 1–20. https://doi.org/10.30953/bhty.v1.13

[39] Zhang, H., X. Chen, X. Lan, H. Jin and Q. Cao, "BTCAS: A Blockchain-Based Thoroughly Cross-Domain Authentication Scheme." *Journal of Information Security and Applications* 55 (2020): 102538.

[40] Wickert, A. K., M. Reif, M. Eichberg, A. Dodhy and M. Mezini, "A Dataset of Parametric Cryptographic Misuses." In 2019 IEEE/ACM 16th International Conference on Mining Software Repositories (MSR), 2019, May, pp. 96–100. IEEE.

6

IoT Benchmark in Industry 5.0

Reshma and Kamal Kant Sharma

Chandigarh University, Punjab, India

CONTENTS

6.1 IoT Overview

IoT has grown to be a marketing trend and general news piece. Beyond exaggeration, IoT appeared as a powerful technique with appliances in numerous domains. IoT has origins in multiple former methods: sensor networks, embedded systems, and pervasive informatics. Many IoT devices are linked mutually to develop specific purpose schemes; in the global network, they are rarely utilized as public access devices.

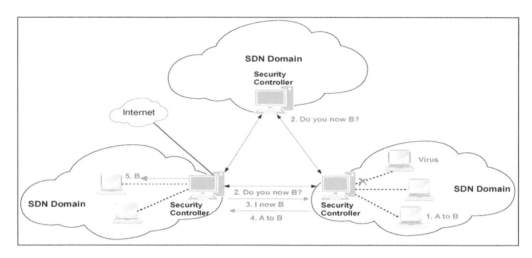

FIGURE 6.1
IoT network architecture [2].

An IoT node is a sensor-contained hardware piece that broadcasts sensed information to users or any other devices over the Internet. IoT nodes embed into industrial equipment, mobile and medical instruments, wireless sensors, and more. Top examples of IoTs are connected smart city, smart industry, smart transport, smart buildings [1], smart energy, smart manufacturing, smart environment monitoring, smart living, smart health, smart food, and water monitoring. Figure 6.1 shows the IoT network architecture. This architecture has a lot of IoT sensors for sensing purposes such as temperature, humidity, pressure, etc. After sensing, these data are transmitted to a cloud server via an IoT gateway. Furthermore, users can access these data through mobile apps and so on.

Due to the accessibility of low cost and smart devices, the IoT network refers to a smart system. IoT devices operate independently with their hearing and transmission abilities. Furthermore, the propagation of IoT provides a lot of benefits but also provides potential threats. An overlooked factor so far is the rise in energy expenditure. IoT nodes are anticipated to always be accessible on other nodes. IoT offers a lot of benefits, including:

- **Locating and tracing abilities:** Customers should be capable of tracking the nodes and locating them in a short amount of time.

- **Ubiquitous information swap:** In IoT where nodes are linked to the Internet and where information is transmitted. Ubiquitous means intelligence. Therefore, intelligence sensors collect information and transmit it using a prearranged input.

- **Enhanced power solution:** Customers should be capable of tracking even the strongest node, and the customer should be capable of obtaining the best result.

- **Data and intelligence management:** IoT does not always require providing commands to the instrument; where the node gives intelligence and information previously it can start working and obtains decisions and discovers solutions based on intelligence.

- **Scalability:** IoT should be the measurability, as with any number of IoT nodes above an extensive network all nodes should distinguish uniquely.

Also, numerous significant IoT problems can be identified. These open problems make it clear that the complexity of Internet design currently needs significant capabilities to alter.

- **Unprotected authorization/authentication:** The administrator usually presents authentication to verify the customer identity, and the authorization utilizes rewriting or modifying the content for that appliance and the consent that the administrator will give.
- **The technology of server:** The number of IoT nodes over the IoT field increases the demand and the number of IoT node replies, moreover increases simultaneously depending entirely on the server where customers use the interface. The server response to the IoT node demand should be made immediately. There must be no delay in responding to the customer.
- **Management of storage:** A massive quantity of information is created. When connected IoT nodes have a massive quantity of multimedia data transmitted, they have big data and other types of inconsistent files where data is held concerning these IoT nodes, these files do not take much space. Still, many of them should be useable as soon as possible.
- **Data management:** As transmission between nodes is completed, more information is created daily between nodes, and there is more information to be transmitted from one location to another. Consideration should be given to whether specific information is transmitted or not.
- **Security:** Provision of security can be challenging as the automation of nodes has increased, which has generated novel security problems.

6.1.1 Essential Characteristics for IoT

An IoT has a fixed framework and comprised of:

- **Enormous scale:** The number of IoT nodes that require handling and connecting will be an order of magnitude higher than the IoT nodes currently linked to the Internet. Of particular importance will be the administration of the information created and its analysis for appliance reasons. It is about content semantics, with content management.
- **Safety:** As customers achieve advantages from IoT, customers should not fail to remember security. As senders and receivers of IoT, they should plan for security. This encompasses the security of customer information and the security of customer welfare: protecting networks, endpoints, and is a universal message that represents generating a measurable safety model.
- **Dynamic changes:** Device status changes drastically, for instance, sleep and wakefulness, connection and disconnection, and content of IoT nodes contain speed and position. Besides, the number of IoT nodes can modify energetically.
- **Heterogeneity:** Various IoT nodes depending on various networks and hardware platforms. They should connect to other nodes via various networks.
- **Connectivity:** This permits IoT to have ease of access and compatibility. Ease of access is obtainable on the network though compatibility, presenting the same capabilities for using and producing content.

FIGURE 6.2
Characteristics of IoT [3].

- **Services related to things:** IoT can present a lot of services related to things within things restrictions. To present services related to things within things limitations, both techniques in the global and data world will alter.

6.1.1.1 Interconnectivity

In terms of IoT, anything could be linked to universal data and contact the IoT basic organizational and physical structures. Essential features are shown in Figure 6.2.

6.2 Benefits of IoT

Currently, every part of business and lifestyle hopes to realize the benefits of IoT. Figure 6.3 shows a list of a few of the benefits that IoT will provide:

- **Technical enhancement:** Similar techniques and data that enhance consumer observation of IoT facts and enhance IoT node usage, and facilitate the most significant advances in technique. IoT opens up a world of actual data performance and field performance.
- **Enhanced consumer engagement:** Recent statistics have the problem of ambiguity and fundamental errors in precision; also, as mentioned, engagement remains are inactive. IoT changes this, attaining a rich and productive engagement, including the spectator.
- **Advanced information compilation:** Today's information compilation undergoes restrictions in plans for practical usage. IoT smashes it down into those gaps and then puts it right where people desire for investigating our planet.
- **Decreased waste:** IoT generates development fields more clearly. Recent statistics provide us with insignificant intelligence; rather, IoT presents actuality data that leads to efficient resource management.

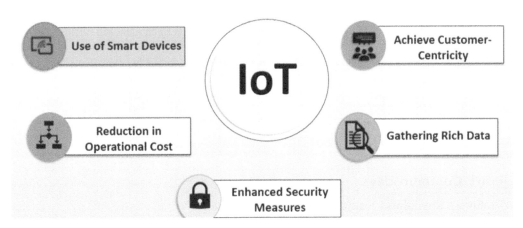

FIGURE 6.3
Advantages of IoT [3].

6.3 IoT Drawbacks

IoT provides an inspiring collection of advantages; it moreover provides an essential collection of challenges. Figure 6.4 shows a list of some of its main problems:

- **Security:** IoT generates an environmental scheme for frequently linked nodes contacting networks. Furthermore, the scheme provides minimal regulation despite safety measurements. It protects consumers from different types of hackers [1].
- **Privacy:** IoT expertise presents private information with complete information, not including the involvement of the consumer.

FIGURE 6.4
Disadvantages of IoT [3].

- **Flexibility:** Consumers are worried regarding the elasticity of the IoT scheme for easy integration. The anxiety is about discovering themselves with too many contradictory or protected source codes.
- **Compliance:** IoT, similar to all other technologies in the trade sector, should obey the rules. Its difficulty creates the problem of compatibility seeming like a daunting challenge when many think that typical software compatibility is a war.
- **Complexity:** A few consumers discover that IoT schemes are complicated in the rule of schema, ordering, and storage provided for their usage of numerous techniques and a massive collection of newly permitted techniques.

6.4 IoT Common Uses

IoT schemes are helpful for many types of appliances:

- Industrial schemes utilize sensors to monitor together with the industry procedures themselves—product excellence—and the condition of the apparatus. A growing number of electric motors, for instance, contain sensors that gather information utilized to forecast future motor breakdowns.
- Smart buildings utilize sensors to discover the positions of persons and the condition of a building. That information can be utilized to regulate ventilation/ air conditioning and lighting schemes to decrease working prices. Smart buildings also utilize sensors to monitor the physical condition of the building.
- Smart cities utilize sensors to monitor persons walking rather than travelling in a vehicle as well as vehicle traffic, and can compile information from smart buildings.
- Vehicles utilize network sensors to monitor vehicle condition and offer enhancement, decrease energy expenditure, and reduce inferior discharges.
- Medical schemes link with a variety of patient monitoring sensors that can be situated at home, in an emergency vehicle, or in a hospital.

There are lots of use cases that assist users in recognizing the needs of the IoT scheme.

6.4.1 Notification System

Messages from IoT devices can be collected and examined. Notices made when certain conditions are met.

- **Sensor network:** The scheme can work definitely as an information collection scheme for the sensor sets.
- **Reactive system:** Study of IoT device sensed information could incite actuators to be accelerated. Users retain a word reactive for schemes that do not execute standard regulatory rules.
- **Analysis system:** Messages from IoT devices are collected and examined, other than in that event, the research is continuing. Research outcomes can be created occasionally.

- **Event latency:** Delays from capturing an event to its receiver cannot be significant for volume-based apps but are significant for online research.
- **Control scheme:** IoT device sensed information is nourishing to regulate the instructions that produce the effects for the actuator. Users may discover the category of non-functional necessities that execute to most IoT schemes. Non-functional necessities in the scheme force non-functional needs on the elements.
- **Buffer volume and event loss rate:** If there are no strict limits on event manufacturing standards, the surroundings can generate multiple events over some time over the scheme. The event loss rate holds the preferred abilities, while buffer volume is a practical need that may directly link to the strength of the elements.
- **Availability and reliability:** As IoT, schemes are dispersed; availability is often utilized to explain dispersed schemes. Reliability may determine across network components instead of total scheme reliability.
- **Throughput and service latency:** Finally, procedures would be executed through maintenance. Users could indicate throughput and delays of maintenance.
- **A lifetime of service:** IoT schemes are always anticipated to contain longevity greater than users anticipate for PC schemes. The life span of a scheme or subdivision of a scheme might be longer than that of an element, mainly if a scheme utilizes passive sensors and different elements.

6.5 IoT Security

Security is ultimately seen as a significant necessity for each kind of computer scheme, containing IoT schemes. However, most IoT schemes are significantly safer than standard Windows/Linux/Mac schemes. IoT safety issues arise from a variety of reasons: insufficient hardware safety elements, inadequately created software with broad limitations of susceptibilities, and different safety creation faults.

Unprotected IoT devices generate safety issues for the whole IoT scheme. Since devices generally contain a lifespan of numerous years, an enormous established foundation of vulnerable IoT nodes would cause safety issues in the future. Unprotected IoT schemes create safety issues across the Internet. IoT nodes abound; unprotected IoT devices are perfectly suitable for a lot of attacks (especially denial-of-service (DoS)). Consumer details are secure from direct theft; however, the IoT needs to be built, thereby low confidentiality information could not be merely utilized for conjecture about high confidentiality information (Figure 6.5).

Example: A German article states that attackers have hacked the safety scheme. They disturbed the management scheme, which prevented the furnace from closing correctly, resulting in serious harm. Therefore, users may know the consequence of the assault earlier, determining a suitable defense.

Challenges: Aside from pricing and availability of IoT nodes everywhere, different safety problems cause continual trouble to the IoT shown in Figure 6.6:

- **Device similarity:** IoT nodes are well homogeneous. These nodes use a similar communication technique and elements. If one scheme or node is vulnerable from susceptibility, numerous others contain a similar problem.

FIGURE 6.5
Encryption roles in IoT [4].

FIGURE 6.6
Security challenges of IoT [5].

- **Unexpected activities:** The vast amount of IoT nodes used and their vast list of empowering techniques denote that these nodes activities in the area may be unexpected. A particular scheme might be well-designed within management systems; however, there are no assurances of how it will communicate with other nodes.

- **Device longevity:** One of the advantages of IoT nodes is long life, but that long life denotes that they can live longer with their node assistance. One can measure the similarity or dissimilarity between this and conventional schemes that have to assist and modernize after a long time, with numerous terminating their utilization. Abandon ware and Orphan nodes do not have similar safety toughness for different schemes because of the emergence of techniques over time.

- **Complex deployment:** One of the primary objectives of IoT is to put superior networks and research where they could not go before. Unexpectedly, this makes it difficult for physical protection of nodes in these areas that are extreme or only accessible with difficulty.

- **Lack of transparency:** Numerous IoT nodes fail to present clarity in terms of their performance. Consumers are unable to see or use their procedures and can only assume the nodes work correctly. They cannot regulate unnecessary activities or information gathering; moreover, when the producer updates the node, it might carry unnecessary activities.

- **No warnings:** One more purpose of IoT is to offer the best performance in the absence of interruption. This initiates consumer consciousness issues. Consumers are not aware of the nodes or recognize when an unfamiliar thing becomes invalid. Safety violations may continue for a long time in the absence of discovery.

6.5.1 Trust for IoT

A trustworthy structure needs to be able to manage people and equipment as consumers, e.g. it needs to transfer hope to people and to be strongly sufficient to be utilized by types of equipment without a DoS attack [6]. The building of hope structures that tackle this necessity will need improvements in fields, for example.

Lightweight public key infrastructures (PKIs) are considered as a basis for confidence management. Improvements are predicted in hierarchy and cross-certification ideas to allow decisions to deal with the scalable necessities.

- Lightweight key management schemes to allow confidence relations to be set up and to share encryption tools using minimal contact and procedure tools, as well as limited resource nature of numerous IoT nodes.

- Data quality is a necessity for numerous IoT-designed schemes where a set of data that describes data about other information can be utilized to offer IoT information reliable estimation.

- Decentralized and self-configuring schemes are substituted by PKI to set up reliability, for instance, a federated identity, peer-to-peer (P2P) network.

- Reliability negotiation is a process that permits two parties to discuss routinely, basically a sequence of reliability tenets, the low condition of reliability needed to allow an act of assistance or a data strip.

- Guaranteed techniques for reliable sites contain protocols, software, hardware, etc.

- Access control to avoid information violations. One instance is usage regulation, which is the procedure of guaranteeing the proper use of specific data according to a pre-defined principle, after which the access to data is allowed [7].

6.5.2 IoT Privacy

Most of the data in an IoT scheme can be private information; there is a need to assist in anonymous and limited private data management. There are many fields where development is needed:

- Cryptographic technologies that allow secured data to be saved and progressed and distributed, without the data being accessible to third parties.

- Securing the secrecy of the area, where the area can be incorporated into things related to human beings.

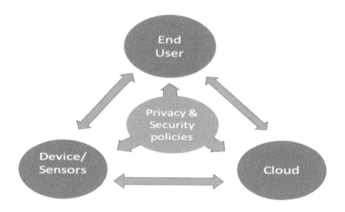

FIGURE 6.7
Framework of IoT.

- Restriction of private data presumption, people may desire to maintain it confidential, by looking at IoT-associated transactions.
- Maintaining data as the neighborhood as is feasible utilizing key management and distributed computing.
- A soft identifier application, where consumer identity can be utilized to make a wide variety of different identities for particular appliances. One soft identifier may not build for a particular theme without exposing unnecessary details, which could lead to a violation of confidentiality (Figure 6.7).

6.6 Functional View of IoT

The IoT system indicates to uniquely discoverable devices with their visual presentations in a structure such as the Internet and IoT problem solving that include many elements, for example:

1. IoT local device communication model (a small distance wireless network can be contacted, for instance, installed on a mobile phone or situated in the surroundings of a consumer, etc.). This model is in responsible for acquiring, monitoring, and transmitting to remote servers for perpetual storage and research.

2. Local research module and observation processing obtained by IoT nodes.

3. Model to communicate with remote IoT nodes, directly on the Internet or possibly by proxy. This model is in charge of observation acquired and transmission to remote servers for perpetual storage and research.

4. Data analytics applications model works on an application server that serves all customers. It takes demands from the web and mobile customers as well as appropriate IoT views as input, implements relevant information executing algorithms, and provides output based on skills that will be provided after to consumers.

5. Consumer interface (mobile or web): measurements of the visual representation in the given circumstances and consumer communication, e.g. description of customer enquiries.

6.7 Application Domains

IoT application areas identified by the IoT European Research Cluster (IERC) based on the inputs of specialists, reports, and studies. Top examples of IoT applications are smart city, smart industry, smart transport, smart buildings, smart energy, smart manufacturing, smart environment monitoring, smart living, smart health, smart food, and water monitoring.

Industrial automation and production stressed from the short-lived manufacturing life cycle and the need for short-term marketing in numerous domains. The next generation of production schemes will be constructed on elasticity and redesigned as an important goal. The new list of IoT appliances presented includes examples of IoT applications in different areas, indicating why IoT is one of the planned technique fashions over the subsequent five years.

6.7.1 Smart Industry

- **Machine-to-machine appliances:** Automatic machine diagnostics and asset management.
- **The liquid level of a tank or vessel:** Monitoring fuel and water levels in storage vessels and wells.
- **Silo volume calculator:** Measure of emptiness and weight of merchandise.
- **Air quality within and around buildings and structures:** Supervising of dangerous gas and oxygen quantities within an industrial process plant that manufactures chemicals to guarantee the safety of workers and merchandise [4,8].
- **The presence of ozone:** Ozone monitoring for the period of dry meat processing in the food industry.
- **Measurement and monitoring of temperature:** Temperature control inside medical refrigerator systems with important ingredients.

6.7.2 Smart Mobility and Transport

1. **Shipment quality:** Supervising of container openings, vibrations, impacts, or any damage for insurance.
2. **NFC payment:** Enable merchants to get credit or debit card payments online by offering a link to a merchant bank or acquirer based in a place or time required to complete the activities of public transportation, museums, galleries, and so on.
3. **Object location:** Search individual objects in large locations such as repositories or ports.
4. **Monitor the activity of fleet vehicles and assets:** Route control of critical assets such as jewels, medical drugs, or hazardous materials.
5. **Non-compatibility discovery of storage:** Alerting of containers' emission of readily combustible materials close to others are holding explosives.
6. **Car management:** Car sharing organizations run the usage of vehicles using smartphones with net connections fixed in each car.
7. **Automatic vehicle diagnosis:** Data gathering from CAN Bus to transmit a real-time alarm to immediate risks or to offer tips to a person who drives a vehicle (Figure 6.8).

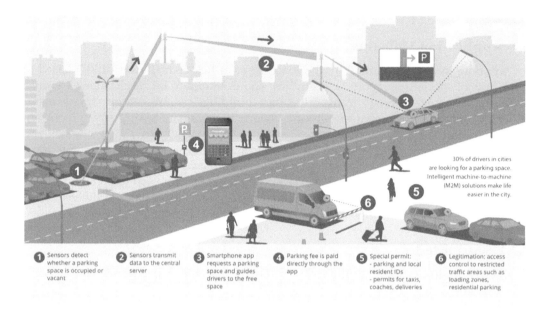

FIGURE 6.8
Applications of IoT [9].

6.7.3 Smart Buildings

- **Liquid availability:** Liquid discovery in data centers, buildings for storing goods, and significant construction underground to avert collapse and the process of slow destruction by chemical action.

- **Access control of perimeter:** Selective restriction of access to protected areas and the discovery of human beings in unauthorized areas.

- **Control of indoor climate:** Measuring and controlling a physical quantity that expresses hot and cold, equipment for producing light, carbon dioxide in fresh air at ppm, and so on.

- **Preserving culture through art:** Condition monitoring within museums and art galleries.

- **Intrusion prevention system:** Discovery of doors and windows openings and breaches to detect intruders, especially into a building with criminal intent.

- **Residential-irrigation:** Smart irrigation and monitoring system.

6.7.4 Smart Energy

- **Installation of photovoltaic:** Supervising and enhancing of operation in solar power stations.

- **Grid modernization:** Power usage supervising and controlling.

- **Wind energy converter:** Supervising and examining the flow of power from a wind energy converter, as well as two-way contact with smart meters for customers to examine usage patterns.

- **Radiation rate:** Shared calculation of radiation rates in surrounding nuclear power plants to create leak warnings.

- **Stream:** Measuring water force that pushes water through pipes in water flow schemes.

6.7.5 Smart Production

- **Composting:** Temperature and humidity management in hay, alfalfa, straw, and so on to deter fungi and various microbiological contaminations.
- **Intelligent management of manufactures:** Controlling stock rotation based on FIFO (First-IN, First-OUT) rules on warehouse shelving and storage to stock replenishment automation.
- **Descendant care:** Controlled breeding in farm animals to guarantee their health and survival.
- **Measurements of toxic gas:** Research of air pollution in farm buildings and the discovery of dangerous gases from stables.
- **Tracking animals:** Locating and identifying animals that graze in open grazing lands or area in the largest stables.
- **Telecommuting:** Providing workers with technology to allow local offices will decrease costs, increase productive capacity, and increase job opportunities while decreasing staff housing, reducing office maintenance and cleaning, and re-moving everyday office travel.
- **Production-line monitoring:** Production-line monitoring and management based on radio-frequency identification (RFID), sensors, video surveillance, remote data sharing, and cloud-based solutions that allow production-line information to change to business-based schemes.

6.7.6 Smart Environment Monitoring

1. **Air contamination:** Decreasing CO_2 emissions from industries, vehicles, and hazardous gases created in agriculture fields [4,10].
2. **Discovery of forest fire:** Flue gases monitoring and monitoring of preventive fire conditions for the identification of warning areas [11–14].
3. **Protection against avalanches and landslides:** Monitoring of the water stored in the soil, density of the Earth and vibrations to discover harmful patterns in earth states.
4. **Wildlife protection:** Tracking calls use GSM/GPS modules to discover and trace wildlife and supply their co-ordinates through SMS.
5. **Premature earthquake discovery:** Distributed control in certain quake areas.
6. **Monitoring oceans and coasts:** Using a variety of sensors combined in aircraft, ships, satellites, etc. for maritime security, fishing vessel tracing, and unsafe oil supplies, etc.
7. **Weather station networks:** Research of meteorological conditions in agricultural land to predict the formation of ice, drought, and air change.

6.7.7 Smart Living

- **Water and energy consumption:** Monitoring energy and water use to get advice on how to save costs and resources.
- **Smart shopping system:** Get advice on where to sell based on consumer pre-ferences, buying behavior, the existence of allergies, or expiration dates.

- **Remote control devices:** Remotely turn on and off devices to prevent accidents and for energy saving.

- **Smart home products:** Transparent LCD display refrigerator that shows what is inside, expired food details, ingredients for a well-stocked kitchen, and all the data obtainable through the app on the smartphone washing machines permit one to control the laundry from far away and work routinely when electricity prices are very low. Smart cooking apps monitor the automatic cleaning function of the oven and permitting adjustable temperature control from a distance.

- **Meteorological station:** Shows conditions of the outside weather, for example, humidity, heat, atmospheric pressure, air velocity, and rainfall using meters capable of transmitting information over long distances.

- **Regular checking of safety procedures and standards:** Baby alarm, an optical tool used to record images and house alarm schemes that are making human beings feeling safer in their everyday lives at home.

- **Gas detector:** Real-time data concerning gas consumption and the state of gas piping can present by linking local gas meters to the Internet protocol (IP) network. Concerning monitoring and assessing water quality, the result could be a diminution in labor and repair prices, enhanced precision and meter readings price reduction, and perhaps gas usage diminution.

6.7.8 Smart Health

- **Activity monitors as support for older persons' physical activity:** Wireless body area network measures movement, critical signals, blurred vision and the cellular unit, gathering, displaying, and storing activity information.

- **Falling detection:** Help for the elderly or disabled who live independently.

- **Pharmacy refrigerators:** The conditions of the regulation within cold storage for storing medicines, vaccines and organic components.

- **Patient monitoring:** Surveillance of patient health status within hospitals and nursing homes.

- **Personal care of sports people:** Vital sign monitoring in high-performance areas and camps. Fitness and health manufacture for these reasons are available, that calculate fitness, steps, weight, blood pressure, and other details.

- **Managing chronic illness:** Patient surveillance programs with full patient details can be obtainable for monitoring chronic illness patients remotely. Admission to a reduced medical facility, low price, and temporary hospital stay can be just a few of the advantages.

- **Hand hygiene approach:** Monitoring scheme by using RFID tags for wristbands with the integration of Bluetooth LE tags at the patient's hand hygiene control, where vibration alerts are transmitted to notify during hand washing. The entire information gathered generates statistics that can utilize tracking patient morbidity in specific health staff [15].

- **Ultraviolet light:** Measuring UV sunlight to alert humans not to be exposed at specific times.

- **Oral care:** A Bluetooth-connected dental brush with an app on the smartphone examines the task of brushing teeth and provides details on the smartphone to obtain confidential information or to show statistics to the dental surgeon.

- **Sleep control:** IoT devices situated across the bed sense usual movements, such as breathing and heartbeat and huge movements caused by tossing and turning at night when asleep, presenting information obtainable via the smartphone app.

6.7.9 Smart Water Monitoring and Smart Food

- **Water leaks:** Discovery of water availability in outside tanks and pressure differences in pipelines.
- **Quality of water:** Analysis of the appropriateness of water in natural flowing watercourses and oceans for all of the animal life present in a particular region or time and the appropriateness for potable usage.
- **Flooding:** Monitoring the variability of flow rates of rivers, reservoirs, and dams.
- **Control the supply chain:** Monitoring storage conditions during supply chain activities and manufacture tracing for tracking reasons.
- **Water management:** Real-time data concerning water condition and water usage can gather by linking a water meter to the IP network.
- **Enhancement of wine excellence:** Monitoring water stored in the soil and trunk width in vineyards to regulate sugar levels in grapes and health of grapevine.
- **Grounds where the game of golf is played (golf course):** Selected irrigation in arid areas to decrease green-water assets.
- **Greenhouses:** Regulate microclimatic level to increase vegetable and fruit manufacture and excellence.
- **In-field water quality monitoring:** Decreasing food spoilage with excellent tracking, statistical management, continuous data acquisition, and crop field management, with excellent management of fertilization, irrigation, and electricity.

6.7.10 Smart City

- **City noise mapping:** Monitoring noise in the concentric zone, including bar areas in real-time.
- **Construction health:** Vibration and conditions of materials monitoring in statues, bridges, monuments, and popular buildings.
- **Traffic jam:** Pedestrian and vehicle level monitoring to enhance walking and driving paths.
- **Safer cities:** Video surveillance system, public address sound systems and fire prevention and control systems.
- **Intelligent lighting:** Smart and weather adaptive lights in raised sources of light on the edge of a road.
- IoT applications differ, and IoT applications work for various consumers. Various categories of consumers have various driving requirements. From an IoT point of view, there are three essential consumer categorizations:
- individual citizens;
- groups of citizens (citizens of a nation, state and city);
- companies (Figure 6.9)

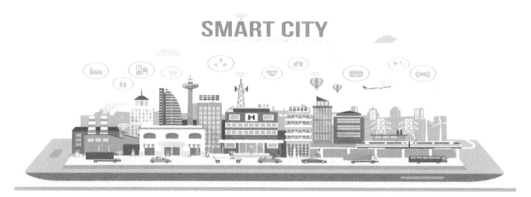

FIGURE 6.9
Smart city [9].

Examples of IoT applications required by individual citizens are such that:

- Increasing their security or the security of members of their family—for instance, remote control for a security alarm system, or activity recognition for older people;
- Facilitating easy-to-perform tasks—for instance: home inventory management reminder.

6.8 Summary

This chapter describes an overview, characteristics, advantages, disadvantages, common uses, security, trust, privacy, and functional view of IoT. Furthermore, we propose application areas of IoT in detail. IoT can promote a functional variety of industrial appliances like logistics, manufacturing, food business, and services. Novel standards, novel trade, competition, and the need to transport nonstop goods are challenges new businesses face nowadays. As a result, a lot of companies rely on Industrial Internet of Things (IIoT), which refers to any performance executed by businesses to model, supervise, and enhance their business processes during insights gathered from thousands of linked machines to assist them in enhancing economical profit. Therefore, the next chapter discusses IIoT in detail.

References

[1] Nia, A. M. and N. K. Jha, "A Comprehensive Study of the Security of Internet-of-Things." *IEEE Transactions on Emerging Topics in Computing* 5 (2017): 586–602.
[2] Olivier, F., G. Carlos and N. Florent, "New Security Architecture for IoT Network." *Procedia Computer Science* 52 (2015): 1028–1033.
[3] https://www.educba.com
[4] Yang, Y., L. Wu, G. Yin, L. Li and H. Zhao, "A Survey on Security and Privacy Issues in Internet-of-Things." *IEEE Internet Things Journal* 4 (2017): 1250–1258.

[5] https://www.netsparker.com

[6] Daud, M., R. Rasiah and M. George, Denial of Service: (DoS) Impact on Sensors, 2018.

[7] Yang, Y., X. Liu and R. H. Deng, Lightweight Break-Glass Access Control System for Healthcare Internet-of-Things. *IEEE Transactions Industrial Informatics* 14 (2018): 3610–3617.

[8] Jamal, H., M. Huzaifa and M. A. Sodunke, *Smart heat stress and toxic gases monitoring instrument with a developed graphical user interface using IoT*, 2019 International Conference on Electrical, Communication, and Computer Engineering (ICECCE), 2019, pp. 1–6. doi: 10.11 09/ICECCE47252.2019.8940738

[9] Kodali, R. K. and S. C. Rajanarayanan, *IoT based indoor air quality monitoring system*, 2019 International Conference on Wireless Communications Signal Processing and Networking (WiSPNET), 2019, pp. 1–5. doi: 10.1109/WiSPNET45539.2019.9032855

[10] Wu, F., T. Wu and M. R. Yuce, *Design and implementation of a wearable sensor network system for IoT-connected safety and health application*, 2019 IEEE 5th World Forum on Internet of Things (WF-IoT), 2019, pp. 87–90. doi: 10.1109/WF-IoT.2019.8767280

[11] Muthukumar, S., W. Sherine Mary and S. Jayanthi, *IoT based air pollution monitoring and control system*, 2018 International Conference on Inventive Research in Computing Applications (ICIRCA), 2018, pp. 1286–1288. doi: 10.1109/ICIRCA.2018.8597240

[12] Prabha B., An IoT based efficient fire supervision monitoring and alerting system, 2019 Third International conference on I-SMAC (IoT in Social, Mobile, Analytics and Cloud) (I-SMAC), 2019, pp. 414-419, doi: 10.1109/I-SMAC47947.2019.9032530

[13] Agarwal K., A. Agarwal and G. Misra, Review and performance analysis on wireless smart home and home automation using IoT, 2019 Third International conference on I-SMAC (IoT in Social, Mobile, Analytics and Cloud) (I-SMAC), 2019, pp. 629–633. doi: 10.1109/I-SMAC4 7947.2019.9032629

[14] Rajurkar C., S. R. S. Prabaharan and S. Muthulakshmi, IoT based water management, 2017 International Conference on Nextgen Electronic Technologies: Silicon to Software (ICNETS2), 2017, pp. 255–259. doi: 10.1109/ICNETS2.2017.8067943

[15] Barman, B. K., S. N. Yadav and S. Kumar, *IoT based smart energy meter for efficient energy utilization in smart grid*, 2018 2nd International Conference on Power, Energy and Environment: Towards Smart Technology (ICEPE), 2018, pp. 1–5. doi: 10.1109/EPETSG.2018.8658501

Section B

Blockchain Applications in Internet of Things (IoT)

7

ASIC-Based Mining System in Blockchain for IoT Environment to Increase Scalability and Security

Bhaskar Dutta[1], Barnita Maity[1], Abhik Banerjee[2], and Tamoghna Mandal[3]

[1]University of Calcutta, Kolkata, India
[2]Netaji Subhash Engineering College, Kolkata, India
[3]National Institute of Technology - Durgapur, West Bengal, India

CONTENTS

DOI: 10.1201/9781003203957-9

7.1 Introduction

Blockchain is a technology that enables decentralized storage of data across multiple notes. Each note stores a copy of the data (also called a ledger) and hence it is also known as distributed ledger technology. Blockchain is immutable, i.e., once any data is stored on the blockchain, it cannot be altered or modified. Because of the intentionally introduced redundancy, it is not feasible to alter any data that is once stored on the blockchain. This makes the system highly secure and users can trust that the data stored is correct.

The process of producing new blocks in a blockchain is called "mining" and the hardware used for this task is termed "miner." The process of mining is associated with solving a cryptographic puzzle that is quite resource intensive. It is possible to use special integrated circuits that are designed solely for the purpose of mining new blocks. Such circuits are known as "ASIC miners." ASIC miners are used only for the purpose of producing new blocks and the design of an ASIC miner will be based on the specifications of the blockchain it will be used in. ASIC miners consume significantly low power and resource when compared with any other alternative for mining blocks in a blockchain.

7.2 Present Problem and Proposed Solution Overview

When a blockchain network is set up, it is important to design how the process of producing new blocks will be carried out. If the network relies on a small number of validator nodes that are responsible for the process of block producing then it centralises the network to some extent. This makes the network highly vulnerable to 51% of attacks [1]. Also, if traditional single board computers are used for the mining process, they are vulnerable to a wide spectrum of security threats [2].

We recommend the use of popular PoW (proof of work) consensus mechanism popularized by Bitcoin because it doesn't rely on centralized validator nodes; rather, every node in the network can participate in the block-producing process. We also strongly recommend the use of ASIC miners because besides being highly efficient, they are also less vulnerable to various security threats. Since ASICs are designed specifically to perform a particular task, the attack surface is significantly low and hence they can't be hacked easily.

7.3 Understanding Blockchain

Blockchain is a way of storing data in a decentralized fashion so that there is not one single point of failure in the entire network. The term "blockchain" is derived from two terms: "block" and "chain." The data is stored in form of small blocks of fixed size and are linked with one another by their hashes. This forms a chain-like structure, hence the term "blockchain" is used (Figure 7.1).

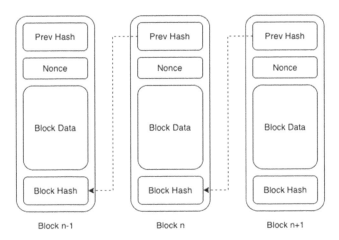

FIGURE 7.1
Structure of a blockchain.

As shown in the diagram, every block in the blockchain consists mostly of

1. Block Data: This contains the main data that is to be stored in the blockchain.
2. Prev Hash: This field stores the block hash of the previous block in the chain.
3. Nonce: Nonce is a special value that is used in the process of block mining.
4. Block Hash: This field stores the combined hash of all the fields defined in the blockchain. The combined hash consists of the Merkle Root hash of all the data, previous hash, and any additional data if present.

Since each block stores the hash of the previous block and the hash of that block depends directly on the prev hash field, if any one block in the chain is changed or altered, the hash of that block will automatically change which will result in any block after that being invalid. Because of this feature of the blockchain, even if the data in one of the blocks is altered, it can be easily detected and the system prevents any such change to take place. For the change to persist, all the successive blocks have to be modified as well which becomes very resource-intensive. The amount of computation needed for such alteration increases exponentially as more and more blocks are added to the chain.

This chain of blocks is also called a ledger. The ledger is distributed among every node present in the network, i.e., every node maintains an up-to-date copy of the ledger. This added redundancy ensures that even if one of the chains in a particular node is compromised or any malicious data is added to it, the other nodes can easily detect it based on a simple majority. Under such cases, the chain with the malicious data will be discarded. Therefore, even if someone wants to add malicious data and corrupt the ledger, at least 51% of the nodes have to be corrupted. This is known as a 51% attack. Given the amount of computation needed for altering a block, such an attack becomes impossible for a blockchain with multiple nodes.

7.3.1 Block Producing/Block Mining

Block producing/block mining is the process of creating new blocks. Since blockchain is based on a decentralized system where every node works parallely and no node has precedence over another, it becomes a challenge to decide which node will produce the next block. Since all nodes have a shared ledger, the block produced by one node must be propagated to the rest of the network. To solve this problem, various algorithms known as consensus algorithms are introduces.

These algorithms define how the nodes attain a "consensus" or agreement among themselves as to which node should be producing the next block. Some of the popular consensus algorithms are:

1. Proof of work (PoW)
2. Proof of stake (PoS)
3. Delegated proof of stake (DPoS)

There are also many other consensus mechanisms [3,4] with new research work conducted for building better and more efficient consensus models. Among the mentioned models, PoW is the most popular [5] as it was used by Bitcoin. In this consensus mechanism, every node competes to solve a cryptographic puzzle, and the first to solve it wins and thus produces the new block. Although this model is more resource incentive as compared to others, it is by far the most decentralized method. In PoS, every node stakes a certain amount of cryptocurrency. The nodes with the highest stake produce new blocks. In DPoS, every node "delegates" or votes for which node to be elected for block producing by staking cryptocurrency. The nodes with the maximum amount of cryptocurrency delegated win and produce blocks. Both PoS and DPoS can be centralised by a single account or stakeholder if that individual holds a very huge amount of cryptocurrency. Such individuals are termed "whales" in the cryptocurrency domain [6].

We will be using the PoW consensus algorithm for our implementation because of its decentralized nature. In the PoW consensus mechanism, a random number called a nonce is generated. A particular value for a nonce is accepted only when the block hash is generated, considering that value statisfies the criteria defined in the genesis block of the blockchain. Every node associated with mining will recursively keep on trying a new random value unless the desired result is achieved.

7.3.1.1 Advantages of Proof of Work Consensus Mechanism

Unlike other consensus mechanisms, in proof of work, all nodes are truly considered equal. There is no predetermined way to find which node in the network will be producing new blocks and it is random; hence, it is not possible to hack the node to produce malicious block data.

7.3.1.2 Hardware-Based Block Producing

ASICs, or application specific integrated circuits, are specially customized integrated circuits which are designed to perform specific tasks. These are highly optimized circuits that can be designed to produce maximum throughput with minimum energy consumption. An ASIC [7] can be used to design miners which are highly efficient in the process of producing new blocks (Figure 7.2).

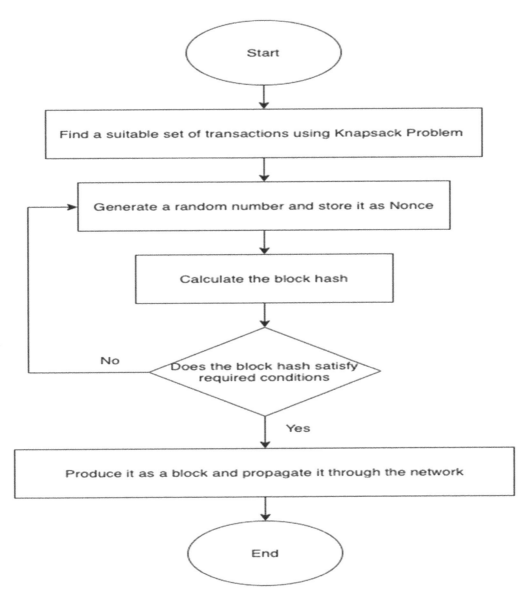

FIGURE 7.2
Flow diagram showing a flow for block producing using proof of work.

7.4 Proposed Model for Implementing IoT in Blockchain

7.4.1 Proposed Node Setup Architecture for Blockchain in IoT

Any setup for IoT devices can be broadly classified into either of the following types.

7.4.1.1 Setup with only IoT Output Devices

In this setup, the IoT device is used to convey a particular output based on the stored data. For this setup, the device only needs access to the data stored in the blockchain. Hence, such a system will only need the computation device for any possible edge computing and reading the data and the IoT device. Traffic lights can be the best example for this system, where the lights are operated based on the data stored (Figure 7.3).

7.4.1.2 Setup with only IoT Input Device

In this setup, there is a sensor that is used to sense and store data in the blockchain. This data will be sent in the form of a transaction. The flow for storing data is described later. In such a setup, there will be a computation device that will connect with other peer nodes in the blockchain, the IoT sensor device, and also a block mining unit. In Figure 7.4, it is displayed as a dedicated ASIC miner because it is the recommended approach; however, it can also be a software running in a light single board computer.

Figure 7.4 shows a high-level overview of such a setup. The data collected from the sensor is passed on to the miner for producing new blocks. The same data is also pushed to other peer nodes. Every node competes for producing new blocks based on the proof-of-work consensus mechanism.

7.4.1.3 Devices with Both IoT Input and Output Devices

In certain cases, a system may have both a sensor device as well as an output device. In such cases, the device will actively record new data as well as generate output based on the recorded and/or edge computation. A dedicated ASIC miner can be used for producing new blocks. Like the previous case, the ASIC can be replaced with software/programs designed for producing new blocks.

Figure 7.5 shows the setup for a node with both sensor and output devices. The data is collected from the sensor and sent to the computation device, which uses the current data

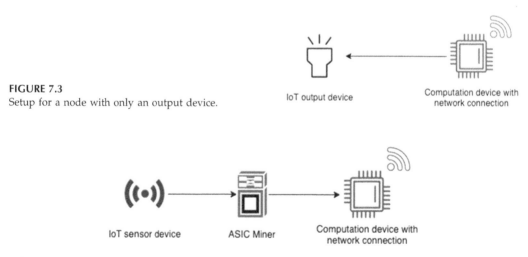

FIGURE 7.3
Setup for a node with only an output device.

IoT output device

Computation device with
network connection

IoT sensor device   ASIC Miner   Computation device with
network connection

FIGURE 7.4
Setup for a node with only sensor device.

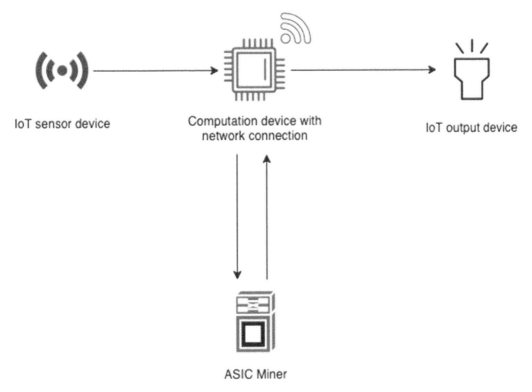

IoT sensor device   Computation device with network connection   IoT output device

ASIC Miner

FIGURE 7.5
Setup for a node with sensor device and output device.

along with records fetched from blockchain to display the required output. The computational unit also forwards the data to the ASIC miner.

7.4.1.4 Devices Only for Mining

It is also possible to have nodes that are present only for mining purposes. Since increasing the number of mining nodes increases the degree of decentralisation and also makes the network more secure, the network may have special nodes which are only present for mining new blocks. These nodes will have only the unit needed for establishing a connection with other peer nodes and an ASIC miner.

Figure 7.6 shows a node that is used only for block-producing or mining. The transaction data is fetched from the peer nodes by the computation unit and sent to the miner. The miner in return performs the mining and sends back the data which is again sent back to the peer nodes.

7.4.2 Block Data Structure and Design

The data is recorded by the blockchain in a format similar to the UTXO model [8], as followed by popular blockchains like Bitcoin. Since the data to be stored in the blockchain will be collected from sensors, certain necessary changes are needed to be made.

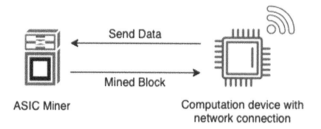

FIGURE 7.6
Only mining node.

7.4.2.1 Understanding the UTXO Model

In the UTXO model, the data stored in the ledger is in form of a list of transactions. When a certain transaction is to be validated, the amount of currency the account is trying to send are compared. Each entry in the ledger has a unique transaction id, the account sending the currency, the account receiving it, and the amount sent in the transaction.

Whenever a new transaction is made, a new transaction record is created. This transaction is not yet confirmed. A list of unconfirmed transactions is maintained. Each transaction also has an associated transaction fee. When a particular transaction is verified and added into a block, the block miner gets this transaction fee. Since each block has a fixed block size, the miner will try to fit in as many transactions as possible so that the total storage size is less than or equal to the block size, and at the same time maximize the fee received. This is a classic example of a Knapsack problem [9], where the transaction fees determine the priority. The higher the specified transaction fees, the higher is the chance of that transaction being confirmed faster.

Figure 7.7 shows the flow where a new transaction is initiated and added to the transaction pool, and then a certain number of transactions are selected and produced into a block and then added to the blockchain.

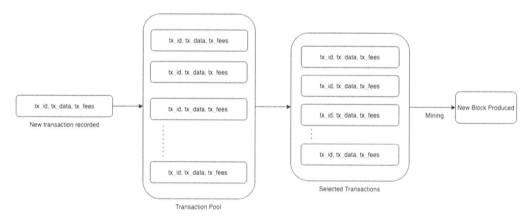

FIGURE 7.7
Flow of transaction selection in UTXO model.

7.4.2.2 Problems with the UTXO Model in IoT

There are certain issues associated with using the UTXO model directly for IoT devices:

1. In a normal transaction, cryptocurrencies or other assets are sent from one account to another but for IoT devices, data will be recorded from sensors and hence it will need a different format. Also, unlike real-world transactions, there is no need for transaction verification or check for double spend.

2. Since transactions are selected from the transaction pool based on the transaction fees, there is a dependence on cryptocurrency for the system to run properly. IoT systems for smart cities will need a blockchain-based solution that can perform efficiently without any dependence on any form of cryptocurrency.

3. If a transaction is not selected from the transaction pool for a long time, that particular transaction is dropped or marked as failed. This means there is an uncertainty that a transaction may fail even if it is valid. This may result in the loss of valuable data when used in an IoT-based system. Hence, there is a need for a mechanism that can guarantee every data is stored.

7.4.2.3 Proposed Model

The proposed model for storing IoT data has the following key differences:

1. Transaction Body: The transaction body will contain a unique transaction id and the sensor data as recorded from the sensor. The sensor will record data at a fixed time interval and create a transaction and pass the recorded data as a transaction body. Since each transaction represents newly recorded data, there is no to or from field. However, if a certain group of data needs role-based access control or encryption, then those data too can be passed in the transaction body. There are no transaction fees involved and every transaction has the same priority.

2. Transaction Pool: When new data is recorded and a new transaction is created, that particular data is propagated to all other connecting nodes. Each node maintains a list of unconfirmed transactions. The list is maintained based on the timestamp of the recorded data in ascending order. If two different records have the same timestamp, they are stored in a queue in the order they are received.

3. Transaction Selection: The transactions will be selected from among the pool of transactions based on timestamp. Since transactions are stored in the pool are also stored in ascending order of timestamp, this means the transactions will be selected in the same order as stored in the pool. The number of transactions that will be selected will depend on the number of transactions that can be stored in a block in proper order.

Since in this model, transactions are selected in the order data is recorded it ensures that each transaction will definitely be mined. Since all transactions contain data from IoT devices, there is no need for a step similar to transaction validation which in return will reduce the processing time. Since the transactions are stored in UTXO format, an algorithm similar to the mining algorithm for Bitcoin can be used.

7.4.3 Mining of New Block

For the block mining process, two types of devices can be used, both having their individual pros and cons.

7.4.3.1 Using an Edge Computation Device Like RaspberryPi

One of the most common implementations would be using SoC (system on chip), which is used for on-edge computation. Such devices have a complete operating system running on them and various forms of applications/programs can be run. Such devices also support complex operations associated with networking and can also perform multi-threading. On the downside, such devices are often more costly and power-consuming. These devices are often a better choice when on-edge operations are to be performed. Such devices are also easy to program since they are capable of compiling and running most programming languages. This also reduces the overhead of developing software for such systems because of the wide range of available libraries.

Such devices can be used for mining as well. If the device is already used for some other computation on edge, multi-threading can be used to run the mining algorithm in a different thread. To mine using these devices, the software has to be designed that can mine new blocks.

Advantages of using such devices for running the miners are:

1. If the IoT device is already running on-edge computation, it will be more efficient to use such devices which can easily run simply to medium level on-edge computation and the mining algorithm can be executed on a parallel thread.
2. It is easier to develop applications for such platforms because they have a full operating system that can compile and run most programming languages.
3. Since such devices are also capable of performing various networking tasks, they can connect with the rest of the blockchain and transmit data without the need for a separate networking module.
4. Updates and bug fixes are easier as the program running on these devices can be easily updated by putting in new code.

Disadvantages of using such devices are:

1. They consume more power and are costly. Such devices often have various modules and features that are not needed for mining and hence if used for mining only will be a waste of resources.
2. The attack surface for any form of cyber-attack increases. Various modules/features that are not used by the system can rather turn into vulnerabilities and can be exploited to get unauthorized access to the devices.
3. Since such devices are made to perform a wide plethora of operations, often they are not efficient in performing specific tasks (in this case, mining new blocks).

7.4.3.2 Using Dedicated ASIC Miners

ASICs, or application specific integrated circuits, are integrated circuits customized for a particular use case, rather than intended for general purpose use. An ASIC miner refers to a device that uses a microprocessor for the sole purpose of mining. The uniqueness of ASIC has revolutionized the way electronics are manufactured. This reduces the die sizes while increasing the density of logic gates per chip.

Advantages of using ASICs are:

1. The small size of an ASIC makes it a high choice for sophisticated systems.
2. In general, an ASIC is optimized to compute just a single function or set of related functions. An ASIC can be used to calculate the block hashes. This would highly increase the hashing rate and hence reduce the time used for mining new blocks.
3. Since they are very application-specific, it is next to impossible to hack such devices as the attack surface is very small.
4. They often consume very little power.

Disadvantages of using such devices are:

1. Making such circuits requires special programming knowledge and equipment.
2. Since they are very specific in function, separate modules are needed for networking.
3. In most cases, they are not easily upgradable.

Since ASIC miners consume less power and have a small form factor, they become the best choice for most IoT setups.

7.4.4 Implementing Pub-Sub Model in Blockchain-Based IoT

Most IoT devices are based on the famous Pub-Sub model [10] where there is a publisher device that is used to publish the data and a subscriber device that subscribes to the publishing device. The subscribing device listens for any data published by the publishing device and performs computation and/or generated output based on those published data.

In a blockchain-based IoT setup, all the IoT devices are considered nodes. The devices that record data will be creating various transactions whereas the devices used for computation and output will only be traversing the blockchain to get the desired data. In order to maintain secure access to data, encryption can be used. Asymmetric encryption with public-private key pairs can be used to encrypt the data. The publishing node can encrypt the data with the public key of the node which is supposed to read the data, and only the node with the correct corresponding private key can decrypt it and get access. In this way, although all the data is stored on the blockchain, only the desired nodes can have access to it.

7.5 ASIC Miners

Mining ASIC technology advanced both in terms of manufacturing technology and in terms of design to achieve greater hash rate, lower power consumption, and lower cost.

ASIC miners perform complex calculations, known as hashing. The technique of cryptographic hash is utilized to achieve this. Hash functions compress a string of arbitrary length to a string of fixed length. The purpose of a hash function is to produce a "fingerprint" of a file, message, or other blocks of data. A hash function must have the following requirements [11–13]:

1. One way property: For any given value h, it is computationally infeasible to find x with H(x) = h.

2. Weak collision resistance: For any given block x, it is computationally infeasible to find y with H(x) = H(y).

3. String collision resistance: For any given block x, it is computationally infeasible to find x and y with H(x) = H(y)

The most common hashing algorithm used in the various blockchains is SHA256. SHA256 produces a 64-character hash. The SHA256 hashing algorithm was developed by National Security Agency (NSA) in 2001.

We recommend the use of double SHA256 or SHA512 because of its capability to convert a larger amount of data into hash more efficiently. The input to the double SHA256 process is a 1,024-bit message, which includes a 32-bit version, a 256-bit hash of the previous block, a 256-bit hash of the Merkle root, a 32 bit timestamp, a 32 bit target, a 32 bit nonce, and 384-bit padding. The 1,024-bit message is split into two 512-bit message parts. Then SHA2561 calculates a hash value of the first 512 bit and SHA2562 computes the hash value of the final 512-bit message.

The main aspects associated are:

1. Padding: The input message is taken and some padding bits are appended to it in order to get it to the desired length. The bits that are used for padding are simply "0" bits with a leading "1" (1000...000). Also, according to the algorithm, padding needs to be done even if it is by 1 bit. If we note M the message to be hashed and l its length in bits where l < 264, then as a first step we create the padded message M, which is M plus right padding.

2. Blocks: M is passed into N blocks of size 512 bits. M1 to Mn and each block is expressed as 16 input sets of size 32 bits.

3. Hash initialization: The initial hash value H0 of length 256 bits is set by taking the first 32 bits of the fractional parts of the sequence roots of the first eight prime numbers.

4. Proof of Work: The miner sets a block and verifies it and will be rewarded for using their CPU power to do so. During the verification of the block, the mines will complete the proof of work which covers all the data of the block, and check whether the hash value of the current block is lesser than the target. Miners compete to complete the proof of work at the earliest.

5. Nonce: The nonce is the central part of this proof of work. The nonce is a random whole number which is a 32-bit field, which is adjusted by miners so that it becomes a valid number to be used for hashing the value of the block. The nonce is the number that can be used only once.

When the collected data is passed through the ASIC miner based on SHA512, a new block will be produced (Table 7.1).

7.6 Comparison Between Various Mining Solutions

TABLE 7.1

Comparisons of Implementation Type

Type	Descriptions	Price	Power Consumption	Implementation in IoT
Antminer	ASIC miners, already available in the market used for mining cryptocurrencies like Bitcoin, Litecoin, etc.	Very High (30,000 INR to 5,00,000 INR approx.)	Very High 3250 W ∓ 5%	Not ideal in IoT because of high price and high-power consumption.
Raspberry Pi	Most common single board computer	Very Low (Approx 5,000 INR for Raspberry Pi 4 B)	6.4 W (For Raspberry Pi 4 B)	Most used device for IoT projects. Low hash rate but efficient for running multiple on-edge applications. More vulnerable to cyber-attacks.
ASIC miner	Custom-designed ASIC miners for block mining. Designed especially for blockchain used for IoT	Based on blockchain parameters can be made at a very low price	Least	Best suited for small-scale blockchain projects like IoT
FPGA [14]	Custom build FPGA unit for mining block in IoT devices	Based on blockchain parameters can be made at a very low price	Least	Easier to build compared to ASIC, power-efficient, and cost-effective.

7.7 Challenges and Scope for Future Work

Using blockchain opens up new opportunities in the field of IoT, but such systems also have their own challenges.

1. Storage Space: Storage space is a big concern when it comes to IoT setups. Because of the nature of the blockchain, the total storage needed to store the entire blockchain always increases. If the entire blockchain is to be stored in each IoT device acting as a node, it won't be feasible because of the limited amount of storage space available for IoT devices. There is a need for a smarter solution wherein it is possible to access the data stored in the blockchain without using excessive storage space.

2. Better Consensus Mechanism: The primary reason for choosing proof of work consensus mechanism of choice was because it ensures complete decentralization and it is possible to design ASIC for this consensus mechanism. However, proof of work is a very resource-intensive algorithm. It's necessary to design a consensus mechanism that is ASIC compatible and also consumes less power.

7.8 Conclusion

Blockchain is a decentralized system that ensures secure data storage and a trustless system. It is possible to build a blockchain-based IoT system where the data collected from various IoT devices are recorded onto the blockchain to prevent any form of tampering. The paper also classifies the various types of nodes that can be present in the system and how they will be interacting with each other. In the end, this paper also discussed how it is possible to use ASIC miners for the process of block mining, which can increase the efficiency of the system.

References

[1] Sayeed, S. and H. Marco-Gisbert, "Assessing Blockchain Consensus and Security Mechanisms against the 51% Attack." *Applied Sciences* 9 (2019): 1788. 10.3390/app9091788

[2] Sainz Raso, Jorge, Sergio Martín, Gabriel Diaz and Manuel Castro, "Security Vulnerabilities in Raspberry Pi–Analysis of the System Weaknesses." *IEEE Consumer Electronics Magazine* 8 (2019): 47–52. 10.1109/MCE.2019.2941347

[3] Fu, X., H. Wang and P. Shi, "A survey of Blockchain consensus algorithms: mechanism, design and applications." *Science China Information Sciences* 64 (2021): 121101. 10.1007/s11432-019-2790-1

[4] Changqiang, Zhang, Cangshuai Wu and Xinyi Wang, "Overview of Blockchain Consensus Mechanism," In Proceedings of the 2020 2nd International Conference on Big Data Engineering (BDE 2020), 2020. Association for Computing Machinery, New York, NY, USA, 7–12. 10.1145/3404512.3404522

[5] Kaur, Sivleen, Sheetal Chaturvedi, Aabha Sharma and Jayaprakash Kar, "A Research Survey on Applications of Consensus Protocols in Blockchain." *Security and Communication Networks* 2021 (2021), Article ID 6693731, 22 pages. 10.1155/2021/6693731

[6] Erich, D., An Examination on The Feasibility of An International Regulated Cryptocurrency, 2018.

[7] Calvão, F., "Crypto-miners: Digital labor and the power of blockchain technology," *Economic Anthropology* 6 (2019) (1): 123–134.

[8] Zahnentferner, J., "Chimeric Ledgers: Translating and Unifying UTXO-based and Account-based Cryptocurrencies." *IACR Cryptol. ePrint Arch.*, 2018 (2018): 262.

[9] Ross, K. W. and D. H. Tsang, "The stochastic knapsack problem." *IEEE Transactions on communications* 37 (1989) (7): 740–747.

[10] Happ, D., N. Karowski, T. Menzel, V. Handziski and A. Wolisz, "Meeting IoT platform requirements with open pub/sub solutions." *Annals of Telecommunications* 72 (2017) (1–2): 41–52.

[11] National Institute of Standards and Technology," Secure Hash Standard (SHA 1)," Federal Information Processing Standard Publication #180-1, 1993.

[12] Korea Telecommunications Technology Association (KTTA)," Hash Function Standard (HAS-160)," TTAS.KO-12.0011, 2000.

[13] Taylor, M. Bedford, "The Evolution of Bitcoin Hardware." *Computer* 50 (2017) (9): 60–61.

[14] Deepakumara, J., Howard M. Heys and R Venkatesan, "FPGA Implementation of MD Hash Algorithm," Electrical and Computer Engineering Canadian Conference on, 2001, 2, pp. 919–924.

8

Usage of Blockchain with Machine Learning for Patient Record Management and Analysis Purposes

Rashmi Sharma, Gunjan Chhabra, and Varun Sapra

University of Petroleum & Energy Studies,
Dehradun, India

CONTENTS

8.1 Introduction

Artificial intelligence is a computationally intensive, intelligent, self-analytic method, which in some ways automatically analyzes and deduces conclusions without manual interventions and interpretations. Almost every human is being directly or indirectly served by these highly specialized systems in their daily lives. To name a few, Google search, Google map, text to speech/speech to text, smart devices, and virtual assistants are such AI-based applications. One can say that no domain is untouched by artificial intelligence now. It is the technology that can predict (based on supervised/unsupervised learning), translate (using natural language processing), navigate the location, and last but not the least identify the fraudulent transactions also (using rule-based algorithms with AI). With all the above-mentioned applications, machine learning, internet of Things, and blockchain are some techniques that are in use. Since the 1950s, artificial intelligence (AI) has come a very long way and nowadays has been applied in almost every aspect of science, engineering, management, and economics. Among all computational streams, the healthcare field is gaining higher penetrance of AI. Additionally, vast developments in medical diagnosis, prognosis, drug design, treatment, and for many

DOI: 10.1201/9781003203957-10

other functionalities are observed. AI is transforming the practice of medication and its administration in entire treatments. Algorithms of AI perform analysis on received data that helps doctors to diagnose patients more accurately, makes predictions about the patient's future health, and recommends better treatments.

Now, the point of concern here is health-related data on which ML will apply. That data should be secure (immutable) and available (sharable) at the time of need (emergency case also). For such cases, a patient has to maintain the record of their health-related issues (documents) in a ready state. For such purposes, blockchain is the easiest way to provide ease to the patient and doctor as well. Hence, here the role of ML in blockchain in medical industries will be described.

The organization of the remaining chapter is as follows: the next section will elaborate the background and preliminaries followed by applications of machine learning and blockchain. Further, with the help of an example, the role of blockchain in healthcare industries will be considered. At the end, the conclusion is followed by references.

8.2 Background and Preliminaries

The meaning of the term *artificial* is either duplicate or substitute of original. Hence, artificial intelligence (AI) is a technique that makes computers observe, contemplate, analyze, and behave like a human brain. AI is the research of analyzing human cognitive intelligence (thinking, learning, selecting, and working) for solving a particular problem. The objective of AI is to imitate human intelligence in computers, such as problem solving, reasoning, and learning. A significant improvement is achieved in many areas of AI. It has several applications in real life, which are often described by using fascinating words, i.e. intelligent, cognitive, smart, and predictive. Specifically it is used in applications where data and information are crucial, i.e. digital health, predictive maintenance, big data analysis, life science, robotics, gaming, expert system, speech recognition, vision system, handwriting recognition, human facial expression recognition, and so on. The issue with AI is that it is a broad concept and often incorrectly interchanged with the other disciplines of AI, i.e. machine learning, cognitive computing, robotics, deep learning, and so on. It includes various sub-domains. Figure 8.1 illustrate some of its major sub-domains that come in its broader umbrella.

Many researchers are working on the amalgamation of AI with many other technologies. Daron Acemoglu and Pascual Restrepo [1] have discussed the role of AI in system automating that helps to reduce the manual labor. T.Y. Wong et al. [2,3] explained the challenges of using deep learning based algorithms in medical field, by taking the data on diabetic retinopathy. Moreover, Hary Surden [4] throws some light onthe impact of AI on all possible fields with the aspect of law. Similarly, Hyun Suk Lee et al. [5] apply AI on physical education with the help of machine learning. In order to make energy-efficient smart buildings under the concept of building management systems, authors Hooman Farzaneh et al. [6] discussed the usage of AI technology with Internet of Things. Similarly, in [7–13], AI techniques have been discussed in many more applications. Moreover, in paper [14], Nisreen Ameena et al. very nicely explain the role of artificial intelligence in online shopping. By applying machine learning on collected customers' data, they explain the relation of data with shopping experience for consumers. Additionally, in the education field, AI plays a vital role to differentiate the understanding level of students

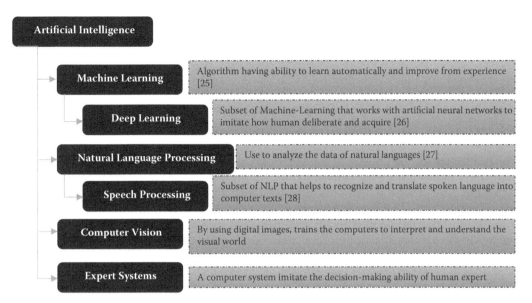

FIGURE 8.1
Sub-domains of artificial intelligence.

by using supervised learning [15–17]. The role of ML and blockchain in the healthcare field will be enlightened in this chapter. Further, the groundwork on healthcare with aspect of AI/ML will be discussed.

In the recent past, AI has spread its aura in the medical field via different routes. Mostly researchers use machine-learning approach to detect health ailments by analyzing the symptoms and their corresponding images taken by X-rays/ultrasounds. In [18,19], the authors have explained the usage of deep learning and pattern recognition with AI in the field of ophthalmology. Similarly, [20] explain the usage of image processing with deep learning to diagnose ear diseases. In [21], authors Guoguang Rong et al. write a review paper on all possible breakthroughs of AI in biomedicine. Likewise, many researchers are working on detection of cancer stages by using the concepts of image processing, neural networking, and deep learning in AI [22–24]. Here, researchers have worked on either affected body organs or diseases where deep learning, image processing, and neural networks or all their combinations are involved (Figure 8.1).

Sub-domains of AI are being used for different autonomus tasks for different fields. In order to automate the system with prediction features, ML and deep learning techniques are very popular. Machine learning is a powerful computational tools that gives effective predictive outcomes and enables us to perceive hidden multi-scale information in a large database. Further, deep learning is another advanced version of ML that helps in overcoming and unlocking much more deeper insights of given data sets irrespective of their complexity and diamensionality.

The overall summary of the mentioned review is that every subdomain of AI is interrelated with each other. Machine learning algorithms are used to train the system with the help of training data. By using neural networks, parameters are fine-tuned upon training data sets, with this training data set analytical, decision-making parameters are adapted and then deep learning algorithms will be able to make a decision on the basis of previous patterns that makes a system expert. Now, here our main motive is to add blockchain in the healthcare field for data storage, security, and retrieval. In the next section, let's have a look at the healthcare system first, then the role of machine learning and blockchain in it.

8.3 About Healthcare Informatics

The healthcare domain is one of the fastest-growing fields of informatics and is creating many new opportunities for healthcare and IT professionals. The past decade has shown incredible growth for the field of healthcare informatics, transforming from old-fashioned paper-based medical records to electronic health records. Of course, the adoption of electronic health records, along with advancements in cloud storage and analytics capabilities, has led to far greater availability (and share-ability) of healthcare data. Healthcare informatics applies principles of computer science and information science to improve the quality of life science research, advancements in public health, health professional education and public health. It is a multidisciplinary field that involves computer, cognitive and social sciences. In healthcare domain, tools and technologies of informatics use to deliver high-end services. Healthcare informatics is a collection of data storage, retrieved and used for providing better health services. Further, this data is available in text, image, signals, and many other formats for future reference and shareability. In the upcoming sections, various AI-based sub-domains are encapsulated to perform various smart tasks for better outcomes. A relatively new specialty in health care, health informatics has transformed how the field operates, which will have larger implications across clinical care, health management, disease prevention, and health policy. While some of those results cannot be foreseen, there are indicators of how health informatics will revolutionize health care across those three disciplines (clinical implications, security implications, and policy implications).

8.3.1 Role of Blockchain and Machine Learning in Health Care

8.3.1.1 Blockchain for securing medical data

Although blockchain was introduced for financial industry transactions, its scope is not limited. This chapter deals with healthcare application of blockchain technology. Healthcare informatics take care of hospital, medical, and human health records at an industrial scale. Human health activities are observed under the supervision of healthcare professionals. Usually it is observed in standard operations that treating facilities do not trust patient's clinical history and previous third-party diagnosis. There are many technical and logical issues, which seem to be apparent and reasonable. When situations are critical and a complete health history remains unknown to a new healthcare worker, it becomes a matter of life and death for the subject under observation. On the other side, if the same history remains available in a (safe, secured) centralized server, the life of a patient becomes easier in multiple dimensions. Blockchain concept applications in healthcare informatics and elsewhere have promising features that may lead us to the next generation of automation. Blockchain features that are uniquely beneficial for this application are immutability, decentralization, enhanced security, distributed ledgers, consensus, and faster real-time updates [25,26]. In [27], the author has shared the research scope in health care with blockchains. We are discussing here the procedure of keeping patients' records secure along with machine learning for data analysis purposes.

In due course of time, we shall be able to see the application of the mentioned technology in the healthcare sector in a broader sense, and it will reduce the manual task by creating a block of patient that includes all previous health details, their updates, along with treating healthcare professional particulars with provided analysis reports, prescriptions, third-party guidance, and so forth. The patients have to deliver only a hash code that shall be provided by the hospital network node (Figure 8.2) or possibly a bio-signature like a fingerprint or retinal or facial scan.

Figure 8.2 explains the aforementioned situation. This figure represents the network of hospitals where two types of nodes are present i.e. validator node and member node. Following are the steps that will be applicable on a network in the presence of blockchain technology:

1. In a blockchain network, a node is just a machine or a computer having required software/hardware specifications to become a node/member of network. Hence, a machine with required specifications registers for node or member of hospital network.

2. After becoming a member, a member node has a right to initiate or receive blocks or data related to patients back and forth to other members (hospitals).

3. After receiving any block, a validator node validates it (assure the correctness of data) and adds it to its centralized ledger (with the help of consensus protocol, validator node take decision).

These above-stated steps are for a blockchain network. Now, a patient's scenario comes in the picture and that is how a patient's data will get created and sent to the concerned hospital or supervisor. Figure 8.3 uses a case diagram to explain the scenario when a patient visits the hospital for the first time. Following are the steps:

Miners (validator node) can initiate/receive and validate transactions

Member node can initiate and receive transactions only

FIGURE 8.2
Hospital network (P2P connection).

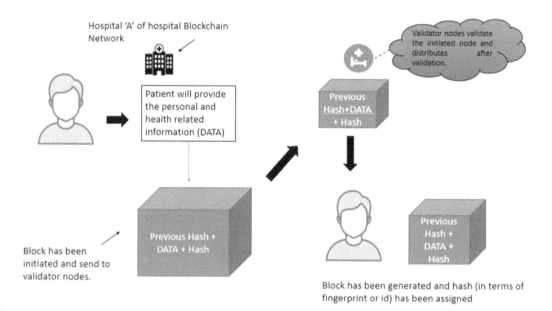

FIGURE 8.3
Block creation at the time of registration.

Fig 1. Patient provides their personal details and symptoms.

Fig 2. Hee prescription accordingly. All these details will be added into the block of subject.

Fig 3. Based on these data, encrypted hash code will generate that shall be provided to the patient for further access.

The previous steps help a patient to create a block and store their health details that will be accessed by authorized persons with the help of a hash code. Now, Figure 8.4 discusses the case when a patient relocates, changes the hospital, or goes for a second opinion (consultation).

Now, the patient has to provide their hash value only to the concerned hospital instead of the entire details. By entering a hash value, all the details including doctors, contact details, and prescription will be available to the hospital representative. Based on the previous history and current symptoms, a new health supervisor will be assigned to the patient. Details of the new supervisor along with the prescription will be mentioned and a new block initiated for validation purposes. This entire process can be paperless and in the case of emergency (traumatic) besides a hash code, facial, retinal, and fingerprints can represent the same personal identity. Afterwards, this new block with a hash value of previous blocks will be added in the chain.

All features of blockchain are taken care while applied in health care. Following are the details of the applied features:

1. **Immutable:** Validators node first validates the data and then adds it to an existing block whose copy is already there with every node. Hence, it makes data indisputable.

2. **Decentralized:** Group of validator nodes are there to maintain the network that makes it decentralized.

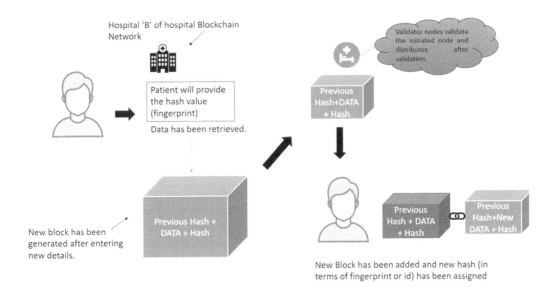

FIGURE 8.4
Addition of new block in a new hospital.

3. **Security:** Encryption algorithms are there to create a hash code for every block that ensures the security of data.

4. **Distributed Ledgers:** Record of patients maintains by all other users/nodes of the network as it is distributed.

5. **Consensus:** As a core of blockchain architecture, consensus helps groups of nodes to make decisions related to blocks. Here, our approach for consensus protocol is a group of health professionals or organizations who define the terms of use.

6. **Faster:** Addition and retrieval of data among network is fast.

With the help of the previous illustration, we discuss the role of blockchain that will reduce the human efforts to maintain the health record and provide security with overall system speed enhancement to that data as well.

The further role of machine learning on data of blocks and on blocks of blockchain will be discussed.

8.3.1.2 Machine Learning in Medical Science

Machine learning is used for data analysis and prediction purposes with the help of supervised or unsupervised learning algorithms. In supervised learning, a model is trained with a labeled data set [28–31], whereas an unsupervised model is trained with an unlabeled data set. Unsupervised learning is used to find the hidden patterns of data, which will help to understand the nature of problem.

These mentioned ML algorithms help a lot in the medical field by maintaining the health record; in the diagnosis of a problem, drug discovery and development and last but not the least for treatment and prediction of disease [21–23]. Here, we will discuss the use of ML within the block and on a complete blockchain.

8.3.1.3 Machine Learning During Patient's Block Designing

We are aware that the block will contain the patient's information along with a hash value. Data of patient can come through any wearable health device or by a traditional method (face-to-face interaction with health professional) and be cross-checked by standard clinical procedures. These data shall feed to the system after cross-validation for diagnosis purposes. Many existing algorithms are there for this purpose [20–24]. Figure 8.5 explains it clearly.

8.3.1.4 Machine Learning Usage on Entire Blockchain

To find the hidden pattern of patients' health from the very beginning to date, unsupervised learning will be used. The first and foremost thing for an IT professional is to understand that there are numerous diagnostic tests and procedures available for disease diagnosis. It is possible to diagnose a disease with many different combinations of diagnostic tests. This analysis will help a health supervisor to check whether a prescribed treatment is helping a patient in their health improvement or not and whether any side effect of the given prescription is there or not. Figure 8.6 explains the concept of ML usage on a patient's entire data.

In Figure 8.6, five blocks are there in a blockchain that contains all transactions (disease related data) of the respective patient. With the help of a previous hash code, traversing of blocks can be done and check the previous data. Now, data of each block will be stored on a common document in a required format and feed that stored data for analysis. As a result, with the help of the following parameters, we can check whether the prescribed treatment is good for the patient or not:

1. Blood pressure
2. Sugar level
3. Any infection
4. Symptoms of existing problem

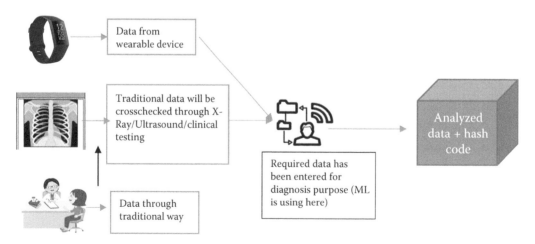

FIGURE 8.5
Machine learning usage in block creation.

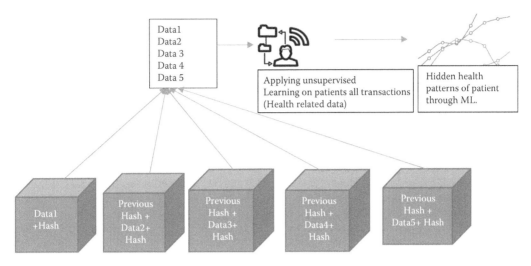

FIGURE 8.6
Machine learning usage on entire blockchain of patient's common disease.

Along with the previously mentioned more health-related parameters can be considered.

The previously stated diagrams demonstrate that relation between blockchain and machine learning is not just within the room, so we can explore it outside also.

8.4 Conclusion and Future Scope

This chapter explains the role of blockchain and machine learning in health care. How blockchain uses it to maintain the record of the patient securely is explained here with the help of case studies or examples. That record is further used for analysis purposes by putting unsupervised learning with it. Additionally, it will be improvised by putting the mathematical model into it along with blockchain implementation with the help of machine learning to determine the health progress report of a patient.

References

[1] Acemoglu, D. and P. Restrepo, *Artificial Intelligence, Automation and Work.* No. w24196. National Bureau of Economic Research, 2018.

[2] He, J. et al., "The practical implementation of artificial intelligence technologies in medicine." *Nature medicine* 25 (1) (2019): 30–36. 10.1038/s41591-018-0307-0

[3] Wong, T. Y. and N. M. Bressler, "Artificial Intelligence With Deep Learning Technology Looks Into Diabetic Retinopathy Screening." *JAMA.* 316 (2016 Dec 13) (22): 2366–2367. 10.1001/jama.2016.17563. PMID: 27898977.

[4] Surden, H., "Artificial Intelligence and Law: An Overview." *Georgia State University Law Review* 35 (June 28, 2019), 2019, U of Colorado Law Legal Studies Research Paper No. 19–22, Available at SSRN: https://ssrn.com/abstract=3411869

[5] Lee, H. S. and J. Lee, "Applying Artificial Intelligence in Physical Education and Future Perspectives," *Sustainability, MDPI, Open Access Journal* 13 (2021) (1): 1–16.

[6] Farzaneh, H., L. Malehmirchegini, A. Bejan, T. Afolabi, A. Mulumba and P. P. Daka, "Artificial Intelligence Evolution in Smart Buildings for Energy Efficiency." *Applied Sciences* 11 (2021) (2): 763. 10.3390/app11020763

[7] Amato, F., et al. "Socially Assistive Robotics Combined with Artificial Intelligence for ADHD." 2021 IEEE 18th Annual Consumer Communications & Networking Conference (CCNC), 2021. IEEE.

[8] Leenen, L., and Meyer, T. "Artificial Intelligence and Big Data Analytics in Support of Cyber Defense." *Research Anthology on Artificial Intelligence Applications in Security*. USA: IGI Global, 2021. 1738–1753.

[9] Ahmad, T., D. Zhang, C. Huang, H. Zhang, N. Dai, Y. Song and H. Chen, "Artificial Intelligence in Sustainable Energy Industry: Status Quo, Challenges and Opportunities." *Journal of Cleaner Production* 289 (2021): 125834, ISSN 0959-6526, 10.1016/j.jclepro.2021.125834

[10] Bakhtiyari, A. N., Z. Wang, L. Wang and H. Zheng, "A Review on Applications of Artificial Intelligence in Modeling and Optimization of Laser Beam Machining." *Optics & Laser Technology* 135 (2021): 106721. 10.1016/j.optlastec.2020.106721

[11] Navale, G. S., et al. "Prediction of Stock Market Using Data Mining and Artificial Intelligence." *International Journal of Computer Applications* 134.12 (2016): 9–11.

[12] Shakya, Subarna, "Analysis of Artificial Intelligence Based Image Classification Techniques." *Journal of Innovative Image Processing (JIIP)* 2.01 (2020): 44–54.

[13] Singh, N., et al. "Implementation and Evaluation of Personalized Intelligent Tutoring System." *International Journal of Innovative Technology and Exploring Engineering(IJITEE)* 8 (2019) (6C): 46–55.

[14] Ameen, N., et al. "Customer Experiences in the Age of Artificial Intelligence." *Computers in Human Behavior* 114 (2021): 106548.

[15] Holmes, Wayne, Maya Bialik and Charles Fadel, *Artificial intelligence in education*. Boston: Center for Curriculum Redesign, 2019.

[16] Devedžić, V., "Web Intelligence and Artificial Intelligence in Education." *Educational Technology & Society* 7.4 (2004): 29–39.

[17] Chen, L., P. Chen and Z. Lin. "Artificial Intelligence in Education: a Review." *IEEE Access* 8 (2020): 75264–75278.

[18] Cheung, C. Y., F. Tang, D. S. W.Ting, G. S. W. Tan and T. Y. Wong. Artificial Intelligence in Diabetic Eye Disease Screening, Asia-Pacific Journal of Ophthalmology: March 2019 – Volume 8 – Issue 2 – p 158–164 doi: 10.22608/APO.201976

[19] Ting, D. S. W., L. R. Pasquale, L. Peng, et al., "Artificial Intelligence and Deep Learning in Ophthalmology." *British Journal of Ophthalmology* 103 (2019): 167–175.

[20] Cha, D, et al. "Automated Diagnosis of Ear Disease Using Ensemble Deep Learning with a Big Otoendoscopy Image Database." *EBioMedicine* 45 (2019): 606–614.

[21] Rong, G., A. Mendez, E. B. Assi, B. Zhao and M. Sawan, *Artificial Intelligence in Healthcare: Review and Prediction Case Studies, Engineering* 6 (2020) (3): 291–301, ISSN 2095-8099, 10.101 6/j.eng.2019.08.015

[22] Ueyama, H., et al., "Application of Artificial Intelligence Using a Convolutional Neural Network for Diagnosis of Early Gastric Cancer Based on Magnifying Endoscopy with Narrow-Band Imaging." *Journal of Gastroenterology and Hepatology* 36.2 (2021): 482–489.

[23] Tran, W. T., et al., "Computational Radiology in Breast Cancer Screening and Diagnosis Using Artificial Intelligence." *Canadian Association of Radiologists Journal* 72.1 (2021): 98–108.

[24] Borrelli, Pablo, et al. "Artificial Intelligence-Based Detection of Lymph Node Metastases by PET/CT Predicts Prostate Cancer-Specific Survival." *Clinical Physiology and Functional Imaging* 41.1 (2021): 62–67.

[25] Cagigas, Diego, et al., "Blockchain for Public Services: A Systematic Literature Review." *IEEE Access* 9 (2021): 13904–13921.

[26] Menon, Nirup M., Byungtae Lee, and Leslie Eldenburg, "Productivity of Information Systems in the Healthcare Industry." *Information Systems Research* 11.1 (2000): 83–92.

[27] Yaqoob, Ibrar, et al., "Blockchain for Healthcare Data Management: Opportunities, Challenges, and Future recommendations." *Neural Computing and Applications* (2021): 1–16.

[28] Dey, Ayon, "Machine Learning Algorithms: A Review." *International Journal of Computer Science and Information Technologies* 7.3 (2016): 1174–1179.

[29] Pouyanfar, Samira, et al. "A survey on Deep Learning: Algorithms, techniques, and applications." *ACM Computing Surveys (CSUR)* 51.5 (2018): 1–36.

[30] Dinesh, P. M., et al. "A Review on Natural Language Processing: Back to Basics." *Innovative Data Communication Technologies and Application*. Singapore: Springer, 2021. pp. 655–661.

[31] Bhatt, S., A. Jain and A. Dev, "Continuous Speech Recognition Technologies—A Review." In: Singh M., Rafat Y. (eds) *Recent Developments in Acoustics. Lecture Notes in Mechanical Engineering*. Singapore: Springer, 2021. 10.1007/978-981-15-5776-7_8

9

IoT-Based Electronic Health Records (EHR) Management System Using Blockchain Technology

Shahid Nazir and Amit Dua

Birla Institute of Technology and Science Pilani-Rajasthan, India

CONTENTS

DOI: 10.1201/9781003203957-11

9.1 Introduction to Blockchain

This section will give a basic introduction of the blockchain technology required for understanding the further implementation details mentioned in the later sections. Blockchain can be thought of as an immutable, distributed ledger that is used to store transactions permanently and securely. This ledger of transactions is stored on every node in the blockchain network. There is no single point of failure because the ledger is distributed and you would have to take down every node in the network to bring the blockchain down, which is almost impossible to do. All blockchain nodes in the network use some algorithm to reach a consensus.

9.1.1 Properties of a Blockchain

9.1.1.1 Provenance

A blockchain is used to store transactions in an immutable and secure way and it can be used to track the origin and ownership of various things like funds, cryptocurrency, assets, etc. The users of the blockchain can trace all the history of these things and verify the authenticity of these things using blockchain in a trustless way. This property of provenance makes blockchain transparent to everyone. This property of blockchain can be used to fight corruption and tax evasion. We can track where the funds are coming from and where they are going to using the blockchain, and we can have transparent movement of funds.

9.1.1.2 Consensus

Since blockchain is a decentralized technology, there is no central authority to maintain and manage the ledger. Also, no one has any reason to trust anyone on the Internet. So, how do we build trust and reach an agreement in the trustless and decentralized network? To solve this problem, we make use of consensus mechanisms that help to achieve agreement among the nodes in a trustless and decentralized way. A consensus mechanism is a set of rules that every node in the blockchain network must follow. Also, every transaction must follow all these rules to be valid. The consensus is driven by incentives that are given to the participants of the network so they don't behave maliciously. The consensus mechanism helps the participants to reach agreement on a global level without any central authority or any middleman. One of the consensus mechanisms that is used by popular blockchain networks is "proof of work," which is used in Bitcoin and Ethereum blockchain networks. Some other consensus mechanisms are "proof of stake," "proof of burn," "proof of activity," etc.

9.1.1.3 Immutability

Blockchain is immutable in a way that once a transaction has been recorded on the blockchain ledger, it is impossible to remove or modify it. That is why every transaction that has been done in the blockchain network and recorded on the ledger is there permanently and no one can change it. This property also makes it secure because no malicious participant in the network will be able to modify anything on the ledger. To change anything on the blockchain they have to change every ledger that is with the majority of the nodes in the network, which is impossible to do. We will see the immutability property of blockchain in detail in subsequent sections.

9.1.1.4 High availability

As there is no central authority, the blockchain network cannot be shut down by anyone and it's running 24×7 every day of the year. To bring down the blockchain network, we would have to shut down every node that is in the network, which is impossible. This property of blockchain makes it available all the time. Also, the blockchain is censorship resistant. No one can stop anyone from accessing the blockchain contents as long as they have Internet access.

9.1.1.5 Finality

Transactions which have been made and stored on the blockchain are permanent and final and cannot be reversed. There is no way to change or remove the transaction once it has been made on the blockchain. This property of blockchain makes blockchain a trustless infrastructure in which everyone can trust the finality and not each other.

9.1.2 Structure of Blockchain

The main data structure used in blockchain technology is linked list. A linked list consists of a chain of nodes that are connected with each other. Each node of the linked list contains a data part and the address part, which contains a reference to the next node and

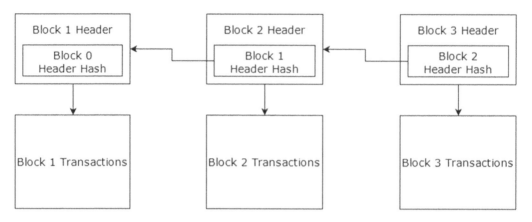

FIGURE 9.1
Blockchain architecture.

this is how the list is connected. In the blockchain technology, addresses are not used for the linking purpose because addresses become obsolete on the Internet i.e. addresses which are relevant to machine X will not be of any use to machine Y. Therefore, in blockchain technology to solve this problem, hashing is used to link the blocks together. Each block in the blockchain consists of a block header and a block body. The block header contains metadata about the block like the timestamp when the block was created, block id, previous block header hash, number of transactions that are in this block, hash of the transactions, etc. The body of the block contains the transactions that are present in that block. The hash that is used to link the nodes is computed on a previous block header and stored as a reference in the next block. This forms a linked list kind of structures with the hashes acting as references to previous blocks. This is why any malicious user cannot change any contents on the blockchain. If they want to change any transaction in any block they must change that block and compute the hash of that block and store that as a reference in the next block and do the same thing for all the following blocks of the blockchain. Also this has to be done for every copy of the ledger for every node in the blockchain network, which is impossible to do, and this is what makes blockchain secure from malicious users. The transactions that are in the blocks contain the sender and receiver addresses and digital signature for that transaction and how much cryptocurrency was transferred if this transaction involved a fund transfer and also contains some other information. The illustration given below shows a simple blockchain (Figure 9.1).

9.1.3 Blockchain Components

9.1.3.1 Ledger

Every node in the network will maintain a ledger, which is a kind of database, and will keep a record of all the transactions in the network. These transactions will maintain the state of the data that is being stored in the blockchain.

9.1.3.2 Smart Contract

A smart contract is just like a contract in the real world, but smart contracts are in digital format. It is a small piece of code that is stored inside a blockchain. Smart contracts execute automatically when certain conditions written in the smart contracts are met. Smart contracts eliminate the need for a trusted third party to make the transaction happen. They are immutable and distributed i.e. once created, the smart contracts cannot be modified or changed. And being distributed means that every node in the network will check and verify the contract's output. Each smart contract invocation is treated as a blockchain transaction. Currently the largest blockchain supporting smart contracts is Ethereum, which was created for the purpose of supporting and running smart contracts. Smart contracts are written in a variety of programming languages, including Golang, Nodejs, Solidity, and others.

9.1.3.3 Peer Network

A peer network is a network of nodes that are involved in the consensus process and they maintain a ledger. They check if the transactions are valid or not and order them and add them to the ledger.

9.1.3.4 Events

Whenever a transaction happens on the blockchain, events are emitted, and the application can register for event notifications. Using those events, we can get all the information about that transaction, like the time of commit or the block number in which this transaction is added or transaction hash, etc.

9.1.3.5 Systems Management

The blockchain network is a distributed system and runs across multiple nodes, so it requires management to create, change, and monitor blockchain components. We can monitor at the system level or application level.

9.1.3.6 Wallet

A wallet is a device that stores the credentials of the user for performing transactions on the blockchain network. In the case of permission-less networks, the credentials are the user's public and private key and in the case of permissioned networks, it is a digital certificate that will be stored in the wallet. It can also perform encryption and signing the transactions. The wallet can be software based or hardware based to store the credentials.

9.1.4 Blockchain Actors

9.1.4.1 Blockchain Architect

The architect designs how the blockchain solution is going to be built. They will figure out how the network itself will be created and what information needs to be stored on the blockchain ledger and what transactions need to happen to access that information. They will also figure out what business logic needs to be embedded onto the blockchain.

9.1.4.2 Blockchain User

The blockchain users will interact with the system and perform business transactions on the blockchain. These users can belong to multiple organizations that are part of the blockchain network. The user is not aware of the blockchain.

9.1.4.3 Blockchain Developer

The developer will take the design or architecture created by the blockchain architect and develop the actual code that will run on the blockchain network itself. They will work on developing the applications and the smart contracts that will be used by users to interact with blockchain.

9.1.4.4 Blockchain Operator

The blockchain operator has special permissions and will manage, monitor, and run the blockchain network. There is a blockchain network operator for every organization in the network.

9.1.4.5 Membership Services

The membership services provide the identities to the users to transact on the blockchain. They will provide a digital certificate to the users and peers so they can communicate using this certificate on the blockchain network.

9.1.4.6 Traditional Processing Platform

These are the systems to which blockchain connects to get some information from or to send information to. These can be traditional data sources or external databases that are also a part of the overall blockchain solution.

9.1.4.7 Blockchain Regulator

The regulator can monitor the operations and transactions happening on the network and they can check if the transactions are legitimate and are compliant with the policies set by the regulator. For monitoring the transactions, they can only have read-only access.

9.1.5 Workings of a Blockchain

The workings of blockchain can be divided into several separate steps, which are discussed below.

Step 1. Some blockchain user creates a transaction and adds their digital signature with it for authentication purposes. The transaction could be transfer of cryptocurrency, invocation of smart contracts or some other information.

Step 2. This requested transaction is broadcasted to the whole network of peer-to-peer nodes.

Step 3. The nodes receive this requested transaction and other transactions from other users. They verify and validate the transactions to check whether the sender is really who they claim to be and if the contents of the transaction are valid. After this, the valid transactions are taken and a block is created from these transactions.

Step 4. After the block is created, the mining process starts, which involves solving a complex mathematical problem. The first node to solve this problem is the

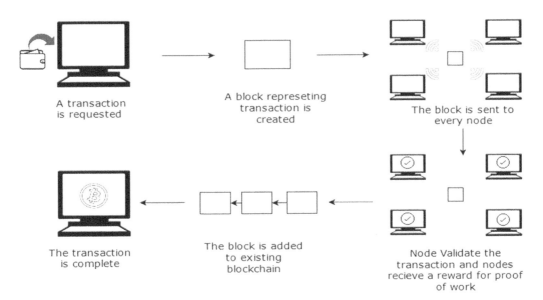

FIGURE 9.2
Workings of a blockchain.

winner and their block gets added to the blockchain and they are rewarded with some incentive generally some amount of cryptocurrency.

Step 5. The transaction is complete after the mining process and it is also reflected in the user's web application.

Following is an illustration of the steps mentioned previously that show the workings of a blockchain from when a transaction is requested to when it is completed (Figure 9.2).

9.2 Types of Blockchains

We've covered the majority of the background information needed to understand blockchain technology, so now we'll look at the many types of blockchains available, each of which has been classified based on its uses and characteristics. Every blockchain network consists of a group of peer-to-peer nodes and every network node holds a copy of the shared immutable ledger, which is updated on a regular basis. Each node has the ability to validate, initiate, or receive transactions, as well as produce blocks. Blockchains can be divided into three categories.

9.2.1 Public Blockchains

Public blockchains are permission-less blockchains. Anyone can participate in a permission-less blockchain as a full node, a lite node, or a community member, for example.

All of the transactions that occur here are totally open i.e. transparent to the public. Anyone in the outside world can observe the transaction. These, unlike private blockchains, are completely decentralized and, as a result, are not governed by a central authority. Many blockchains currently require tokens (the blockchain's cryptocurrency) to encourage and reward their users due to the lack of a central body. One of the goals of introducing a token is to build confidence in the blockchain network. Because it is almost impossible to modify or amend data once it has been validated and uploaded to the blockchain, data on a public blockchain is secure. Ethereum and Bitcoin are well-known examples of public blockchains.

9.2.2 Private Blockchains

A private blockchain is a permissioned blockchain operating in a closed network only. Permissioned blockchains are a private network to which only a select few people have access. The advantage of permissioned blockchain over permission-less is that they provide access control to the blockchain. The transactions performed by the participants who are part of this blockchain are kept concealed from the rest of the world. These blockchains are frequently utilized within businesses when information is deemed too sensitive to be shared with the general public. You must first get authorization from the blockchain's central authority to validate transactions made on these types of blockchains. Hyperledger Fabric and Ripple are examples of private blockchains.

9.2.3 Consortium Blockchains

A consortium blockchain is a semi-decentralized type of blockchain in which a blockchain network is managed by many organizations. In this sort of blockchain, more than one organization can function as a node, exchanging information or mining. Banks, government agencies, and other institutions frequently employ consortium blockchains. Energy Web Foundation, R3, and other consortium blockchains are examples.

9.3 Hyperledger Fabric

The Linux Foundations built Hyperledger Fabric, an open-source technology. Hyperledger Fabric is designed to serve as a basis for building modular products and solutions (Hyperledger Fabric – Hyperledger, n.d.). It's a solution for establishing and managing permissioned blockchains that are modular and extensible [2]. This platform will be heavily utilized to meet our permissioned blockchain demands. This is a simple platform with a variety of modules and functions that may be leveraged without having to re-implement them in the application. As a result, such features can be easily reused, and current functionality can be easily integrated. It also allows us to designate network participants with entirely distinct levels of authority and functioning.

9.3.1 Hyperledger Fabric Architecture

The hyperledger fabric architecture consists of some extra nodes and components than the regular permissionless blockchain. Hyperledger fabric consists of a modular architecture i.e. every component is pluggable. The fabric network consists of following components and nodes as shown in the figure and also a brief explanation of each of the components and the nodes which are a part of hyperledger fabric architecture is given below (Figure 9.3).

9.3.1.1 Membership Service

The membership service will provide a notion of identity for the users who are going to be transacting on the blockchain. This identity is going to be a digital certificate and users will use this digital certificate to sign their transactions and submit to the blockchain. The benefit of signing this transaction is to authenticate to the blockchain that the user is a legitimate user and it also ensures that the users get right access privileges for transactions they are performing on the network. The digital certificate is provided by a certificate authority. The certificate authority is a pluggable module, and it can be any external certificate authority or the default fabric certificate authority, which is implemented by a fabric network. All of this uses a public-key-based infrastructure.

9.3.1.2 Fabric SDK

The SDK is used to interact with the blockchain network and is available in multiple languages like NodeJS, Python, and Java. We can use any of these SDKs to perform transactions on the blockchain.

9.3.1.3 Endorsing Peer

The endorsing peer executes the proposed transaction and does not update the ledger. The endorsing peer collects the data elements that were written and read during the execution of the transaction. These data elements that were read and written are known as RW sets and then the endorser digitally signs them and sends them back to the client

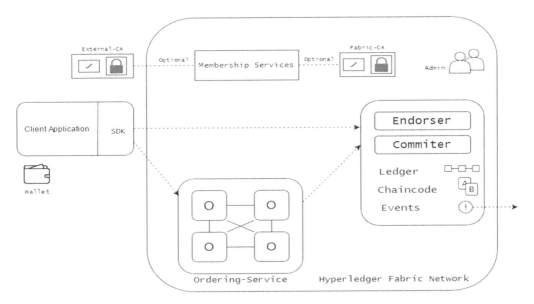

FIGURE 9.3
Hyperledger fabric architecture.

application. The endorsing peer can be a committing peer also or only an endorser. The endorsing peer must hold a chaincode but it may or may not have a ledger.

9.3.1.4 Committing Peer

The committing peer commits the block received from the order into its ledger after validating the transactions in the block. Once the committer validates all the transactions in the received block, it is added to the committing peer's ledger and the world state is also updated with the Write data from the RW Set received in the transactions. The committing peer can also be an endorser i.e. it can have smart contracts as well.

9.3.1.5 Ordering Peer

The main function of the ordering peer is to receive the endorsed transactions and to order them into a block. These ordered blocks are then delivered to the committing peers by the ordering peer. The main function of the ordering service is to provide a totally ordered set of transactions to the nodes. The ordering service itself can be decentralized and may consist of many nodes. These nodes do not have smart contracts and ledgers.

9.3.2 Chaincodes

Chaincodes are basically smart contracts in a hyperledger. They execute when certain conditions written in them are met. The chaincodes are written in various languages like Go, node.js, or Java. The chaincodes are deployed on the endorsing peers and once deployed they can be invoked using the transactions only and using the transactions the ledger state is updated.

9.3.3 Transaction Flow

Consensus is achieved using the following transaction flow: Endorse transaction then order transaction then validate transaction. The client application is going to submit the transaction to a few peer nodes and these peers are going to execute this transaction and get the output and add their signature to the output. After that, these signed outputs will be sent back to the client application. The client application after receiving these endorsements will check if all the outputs are the same and it will confirm that this is a valid transaction. This is the endorsement part. Once the transaction is found to be valid after the endorsement step, then it is submitted for ordering. The ordering service will ensure that the transactions are totally ordered across all the nodes. After that, there is a validation step that ensures that two transactions that are trying to modify the same state should not be executed simultaneously. The transaction that will be executed out of these two transactions will depend on the ordering service.

  The detailed explanation of the transaction flow of hyperledger fabric is given below. To explain the transaction flow, I will consider a hyperledger fabric network that has three endorsing peers E1, E2, and E3 and two committing peers C4 and C5. E1, E2, and E3 will also be committing the transaction i.e. they will also be the committers. Each of them has a ledger to store transactions. Each of the endorsing peers have chaincodes "A" and "B" installed on them. Also, there is an endorsement policy "P" defined which says "E1, E2, and E3 must sign the transaction." C4 and C5 are not part of the policy and they only have to commit the transaction. There is also an ordering service in the network that has four orderer nodes "O."

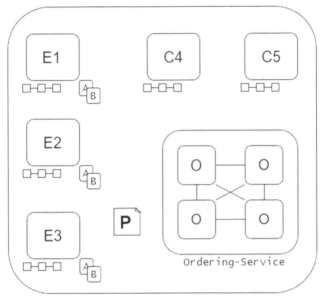

FIGURE 9.4
Fabric network to explain transaction flow.

The following figure shows the network that I will use to explain the transaction flow (Figure 9.4).

9.3.3.1 Step 1: Propose Transaction

The client application or a user is going to propose a transaction. In this proposal, they are going to say what their identity is and what function of which smart contract they want to invoke, and they will also give the inputs. The client application will send this to the endorsing peers E1, E2, and E3.

9.3.3.2 Step 2: Execute Proposed Transaction

Now all endorsers E0, E1, and E2 in the figure will execute the requested transaction but they will not update the ledger. They will capture which data elements were read and written which are called RW sets and these are captured by every endorser node to which transaction was proposed. These read-write sets will be signed by the respective endorsers and will also be encrypted, which will now flow in the fabric (Figure 9.5 and 9.6).

9.3.3.3 Step 3: Proposal Response

The endorsers will now send these read-write sets to the client application in an asynchronous manner. The client will collect these endorsements from the endorsers to which it has proposed the transaction (Figure 9.7).

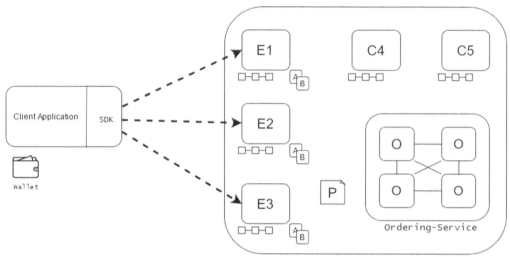

FIGURE 9.5
Transaction proposal.

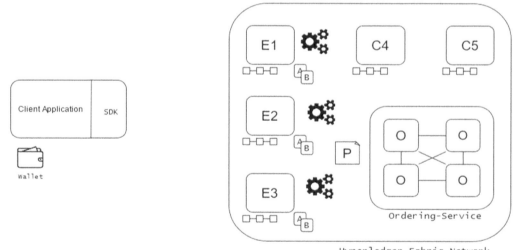

FIGURE 9.6
Transaction execution.

9.3.3.4 Step 4: Order Transaction

After receiving the sufficient responses from the endorsers that satisfy the endorsement policy, it will send the transaction that contains responses to the order. In our case, if the application receives responses from E1, E2, and E3, it will submit those responses in the form of a transaction to the ordering service. These submissions of transactions to the ordering service will be happening from multiple client applications and users simultaneously across the network. The ordering service will order these transactions and it will make sure

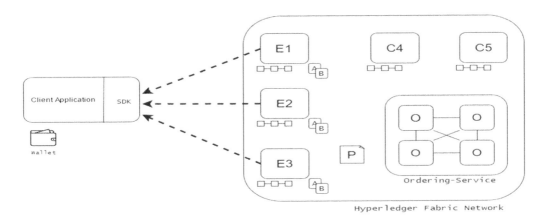

FIGURE 9.7
Client application receives responses.

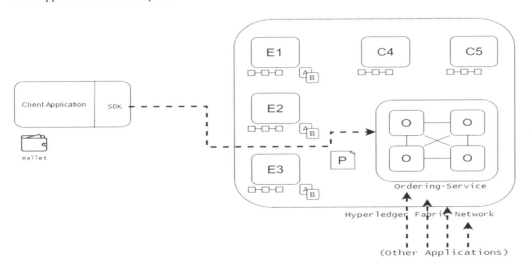

FIGURE 9.8
Send transaction for ordering.

that everyone across the network sees the same order. There are different ordering algorithms available for ordering the transactions and it depends on the ordering service which one it implements. Some of the algorithms are FIFO, KAFKA, and Solo (Figure 9.8).

9.3.3.5 Step 5: Deliver Transaction

After the ordering service orders the transactions, it forms a block of these ordered transactions and delivers it to the committing peers in the network i.e. to E1, E2, E3, C4, and C5 in our case (Figure 9.9).

9.3.3.6 Step 6: Validate Transaction

All the peers have received the blocks. The endorsement policy is checked by each peer. Also, make sure that read-write sets are always valid in the present world state. Validated

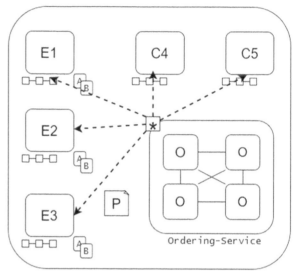

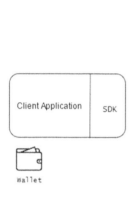

FIGURE 9.9
Deliver block to all peer nodes.

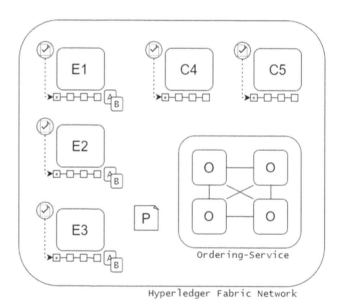

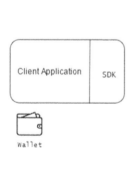

FIGURE 9.10
New block is added to blockchain.

transactions are saved to the ledger and then committed to the world state. The invalid is likewise kept in the ledger; however, the world state is not updated (Figure 9.10).

9.3.3.7 Step 7: Notify Transaction

The committing peers after adding the block to their ledger will emit events for the success or failure of transactions. These events can be listened to by the applications if

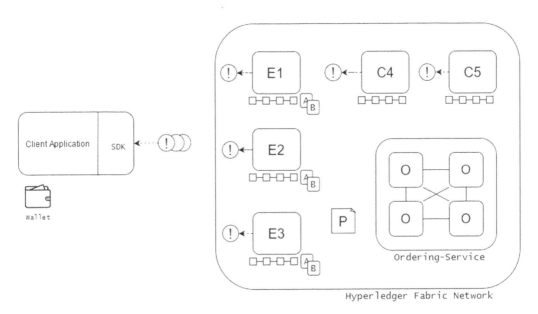

FIGURE 9.11
Send notifications to the client application after block addition.

they register to be notified. They can get events from multiple peers for the proposed transaction. In the smart contracts, we can embed events when the state of variables changes (Figure 9.11).

9.4 Challenges in Current Electronic Health Record (EHR) Systems

Before the advancement in current technology, the healthcare industry relied on a paper-based system for storing medical records that was a handwritten system. The problem with this system was that it was inefficient, unreliable, and not secure. It had a severe problem in terms of accessibility, and it could only be used by one person at a time.

With the advancement in technology came EHR systems that are being used to store health records of the patients in a digital format. However, these EHR systems have concerns regarding the medical record security and privacy, user data ownership, and data integrity. The majority of EHR data sharing presently occurs by fax or mail, which can cause considerable delays in patient care and this is due to the lack of a secure and trustworthy medical records sharing system. There are multiple EHR systems in use at various institutions, each with its own set of terminology, technical capabilities, and functional capabilities; hence, there is no globally established standard. Accessing fragmented patient data across several EHRs is still a challenge. A patient must go through a lengthy and laborious process in order to view all of their medical records.

Breach of data in the medical industry is also a serious concern. There are no alerts triggered for unauthorized access of patient information, and it is not known what information was accessed.

9.4.1 Challenges in Current Health Monitoring Systems

While an increasing number of elderly require long-term care, many still desire to stay independent and active as long as possible and reside in their own homes. At home, the patients cannot be treated and taken care of due to the lack of necessary medical equipment. In hospitals, continuous monitoring of patients' essential vital signs is a critical task. This is generally done nowadays with the use of several cabled sensors affixed to the patient and connected to bedside monitors. Due to this the patients have to be fixed with devices near their beds to monitor their vitals.

9.5 Motivation for Introducing Blockchain into the System

Given the aforementioned issues, it is reasonable to seek a platform that may aid in the transformation of the healthcare sector to one that is more patient-centric. Blockchain can provide such a platform that is both secure and transparent, as well as provide data integrity for patient medical records without third-party intervention. By using blockchain, the patient can keep track of who accessed their medical records and every access to the medical records can be stored in the blockchain permanently in the form of transactions. Also, to provide access control, permissioned blockchain can be used so that only certain identifiable participants can access the system. In our case, we want only the patient and the doctor who has been given access by the patient to view the EHRs.

9.6 Proposed IoT-Based Blockchain EHR System

We propose an IoT-based permissioned blockchain system that could be utilized in the healthcare sector to deploy blockchain technology for EHR and remote health monitoring. The goal of our proposed system is to store the medical records and patient's vitals data in a secure way and to give access to those medical records to those only whom the patient allows so that our system is patient-centric. Furthermore, by leveraging cloud storage for medical records, this framework tackles the scalability issue that blockchain technology presents in general. Introduction of IoT sensors in this system helps in gathering the patient's vital information in a remote and secure way. As this system is built on permissioned blockchain, therefore, the access to the blockchain can be controlled and the patient can control who can access their medical records. Also, this system will be secure and scalable because of the use of blockchain technology.

9.7 Technical Details

The EHR blockchain network consists of peer nodes (hospitals, clinics, etc.), hospital administrators, a network administrator, and a cloud server for storing data. Below is a description of the various entities that make up the system.

1. **Patient:** The patient will be uploading their EHRs to the cloud or they can grant access for a certain period to his/her EHRs to the doctors and after that, they can also revoke access to his/her EHRs. All these accesses to their EHRs will be stored in the blockchain in the form of transactions.

2. **Hospitals:** The network administrator registers the hospitals, and they are added to the network as organizations. Every hospital in the network must provide a peer node that needs to set up a hyperledger fabric peer, a CouchDB instance for on-chain metadata management, and a web application to communicate with the chaincode and EHR. Also, each hospital has an administrator who is responsible for adding and removing doctors and patients from the system.

3. **Network Administrator:** A network administrator is a trustworthy individual who is in charge of the overall system's management as well as the registration and addition of hospitals to the system.

4. **Cloud Server:** A cloud server is a reliable infrastructure with significant computing speed and storage (cloud storage). To enable safe data sharing and storage resources, a cloud server is used to store and administer the patient's EHRs. The cloud server will be receiving the EHR data from the patient after a transaction for the addition of data has been made in the blockchain.

5. **Mobile Device (MD):** A simple mobile device that is used to connect the patient's sensors with the cloud. This will take the data from the sensors and do some computation and send that data for storage to the cloud. Every patient who is fixed with the sensors will have a mobile device.

9.8 Workings of the System

9.8.1 Medical Records Storage

In the proposed system, a blockchain node will be provided by each hospital integrated with its own EHR system that will form the blockchain network. Patients and doctors will use a web-based interface to upload EHR records and to access them. The medical records will be stored off-chain in cloud-based storage and only the metadata about those medical records will be stored on the chain. The metadata that will be stored in the blockchain for each patient will include patient ID, patient name, hash of the emergency records of the patient, and hash of the encrypted records of the patient. The patient's EHR data will be stored in two layers, differentiating their regular and emergency health records. The regular layer data will be stored directly in the cloud storage, while emergency records will be encrypted first and then stored in the cloud. To secure the EHR data of the patient, the system employs public-key-based asymmetric encryption and also uses digital signatures.

The patient, when visiting the doctor, can add them to the list of doctors who can view the patient's regular records. The patient can also give access to their emergency records to the doctor. And after a visit to the doctor, the patent will remove the doctor's access to their records. The doctor on the other hand can upload the prescription for that patient which will be added to the patient's EHR data and will be stored as a transaction in the blockchain which will prevent prescription fraud.

Figure 9.12 shows the system model of the proposed electronic health record (EHR) management system using blockchain.

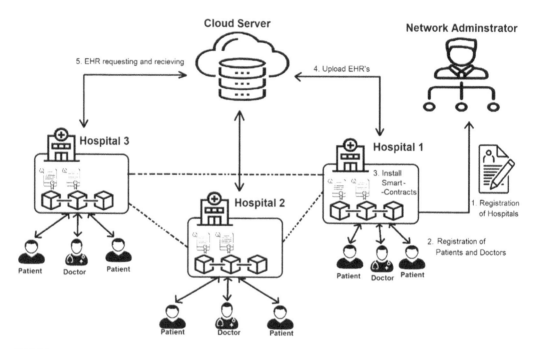

FIGURE 9.12
EHR management system model.

9.8.2 Health Monitoring System Using IoT and Cloud Services

This is the second module of our system that is used to monitor vitals of the patients and then send them to the cloud for storage. The workings of this module are given below.

The patient will be fitted with the IoT sensors like (ECG, blood pressure monitor, blood glucose meter, body temperature thermometer, etc.) and these will be used to collect biological data. The IoT sensors which are fitted to the patients will be connected to their mobile device wirelessly and this will be used to process the data and generate aggregate (like avg, min, max, etc.) of the whole data for fixed period intervals and then sent to the cloud. The mobile device will be receiving data from IoT sensors in real time and if it receives any values that are below or above the set threshold for that specific value, then it will send it is immediately sent to the cloud and the cloud will send a notification to the doctor. The doctor can only view this data if they are allowed to by the patient and can only do so by transacting on the blockchain. This way, the patient can see who accessed their data and they can control the access to their data and not the doctor. After the patient gets well, they can remove the access of the particular doctor to their sensor's biological data so that the doctor can no longer view it.

The following figure shows the system model of the proposed health monitoring system (Figure 9.13).

9.8.3 Algorithms

This section will provide a description of all the algorithms used in the system. These algorithms are used to update or access the blockchain ledger.

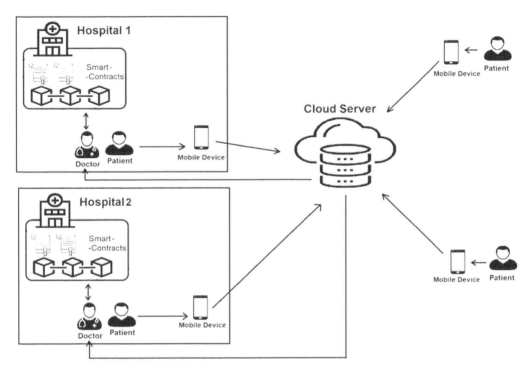

FIGURE 9.13
IoT health monitoring system model.

ALGORITHM 1   ADD EHR IN THE BLOCKCHAIN

Inputs: PID, P-name, Hemergency, Hencrypted
Output: EHR is stored in the blockchain

1. if number of inputs is 4 then
2. Create an empty EHR structure "EHR_sample"
3. EHR_Sample.P_ID = PID
4. EHR_Sample.P_Name = P_name
5. EHR_Sample.hash_encrypted = Hemergency
6. EHR_Sample.hash_emergency = Hencrypted
7. PutState(EHR_Sample.P_ID, EHR_Sample)
8. else
9. return Error;
10. end

ALGORITHM 2   UPDATE HASH OF EMERGENCY RECORDS

Inputs: PID, Hemergency
Output: Hemergency of PatientPID is updated

1. if number of inputs is 2 then
2. Create an empty EHR structure "EHR_sample"
3. EHR_sample = GetState(PID)
4. EHR_Sample.hash_emergency = Hemergency
5. PutState(EHR_Sample.P_ID, EHR_Sample)
8. else
9. return Error;
10. end

ALGORITHM 3   UPDATE HASH OF ENCRYPTED RECORDS

Inputs: PID, Hencrypted
Output: Hencrypted of PatientPID is updated

1. if number of inputs is 2 then
2. Create an empty EHR structure "EHR_sample"
3. EHR_sample = GetState(PID)
4. EHR_Sample.hash_encrypted = Hencrypted
5. PutState(EHR_Sample.P_ID, EHR_Sample)
8. else
9. return Error;
10. end

ALGORITHM 4   GET HASH OF ENCRYPTED RECORDS

Inputs: PID
Output: Hencrypted of PatientPID

1. if number of inputs is 1 then
2. Create an empty EHR structure "EHR_sample"
3. EHR_sample = GetState(PID)
4. return(EHR_Sample.hash_encrypted)
5. else

6. return Error;

7. end

ALGORITHM 5   GET HASH OF EMERGENCY RECORDS

Inputs: PID
Output: Hemergency of PatientPID

1. if number of inputs is 1 then

2. Create an empty EHR structure "EHR_sample"

3. EHR_sample = GetState(PID)

4. return(EHR_Sample.hash_emergency)

5. else

6. return Error;

7. end

ALGORITHM 6   GET NAME OF PATIENT PID

Inputs: PID
Output: Name of PatientPID

1. if number of inputs is 1 then

2. Create an empty EHR structure "EHR_sample"

3. EHR_sample = GetState(PID)

4. return(EHR_Sample.P_name)

5. else

6. return Error;

7. end

9.9 Implementation Details

This part will tie together all of the previous sections and offer precise processes that the system will perform while utilizing all of them.

- At first, hyperledger fabric will be installed with all required docker images, including organization, peer, CouchDB, orderer, and other modules. The fabric network and the above components will run in the background.

- Then the chaincodes will be installed on the peer nodes.
- Once the hyperledger network is up and running, then we will start the front end i.e. the web application which the doctors and patients will use to interact with blockchain.
- The web application has three users: administrator, doctors, and patients.
- The administrator can add the new hospitals in the blockchain network or remove them and for each hospital they will add a hospital administrator (Figure 9.14).
- Every hospital will have one administrator who will be managing the addition or removal of doctors from a particular hospital.
- The patients, after registering in the web application, can upload their medical records that will be stored in the off-chain storage i.e. cloud in our case (Figure 9.15).
- The medical records can be uploaded in pdf or jpeg and stored format.
- The patients can give access to or revoke access from a particular doctor from a particular hospital for any of their specific medical records (Figure 9.16).
- The doctors can view the patients' records for which they have been given access to (Figure 9.17).
- The access to any of the patient's vitals data can also be given to the doctor and revoked also (Figure 9.18).

9.10 Conclusion

We have proposed an IoT-based EHR management system that solves a lot of problems, which the current electronic health record system suffers from regarding security and accessibility. In this chapter, we saw how blockchain technology can be used to overcome those problems. We also saw problems that can be solved by remotely sensing patient's vitals using IoT sensors and how that data can be stored in the cloud and can be accessed by blockchain transactions only. In the first section, a comprehensive explanation of all the fundamentals needed to comprehend how blockchain technology works was provided. Then a brief explanation of different types of blockchains was provided and it was found out that permissioned blockchain was relevant for our system. We also chose a hyperledger fabric network, which is a permissioned blockchain network to implement our system. We then gave a description of different components and actors involved in the fabric network and the transaction flow in fabric.

In the next section, we talked about the challenges in current EHR and current health monitoring systems and motivation for this system. Also, a detailed overview about our system was provided. Then we talked about the architecture of our system and the workings of our system. The next section provided the pseudocode of the algorithms used in the system. Finally, in the last section, to tie everything together, a detailed description of how the system would work and its implementation details were provided.

Security and controlling access to the medical records are the most prominent challenges in the existing electronic health record system. So, this chapter presents a system that can transform the healthcare sector and can be used to solve a lot of the mentioned problems present in the healthcare sector. The whole system is built around hyperledger fabric, which is a permissioned blockchain and the proposed system is completely patient-centric. [1,3–20]

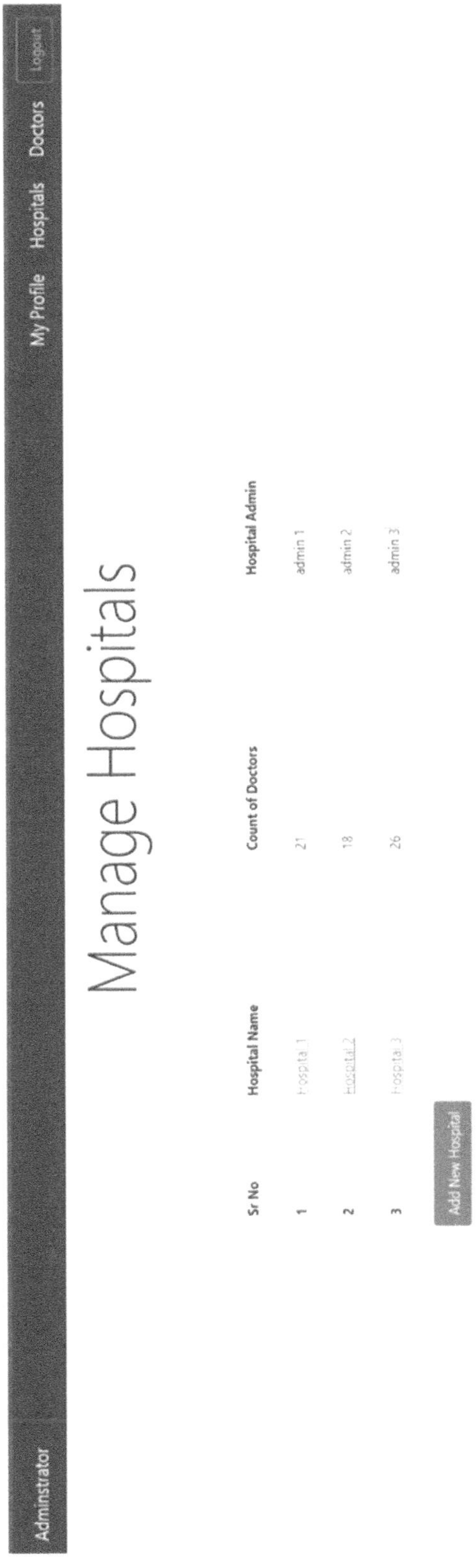

FIGURE 9.14
Network administrator.

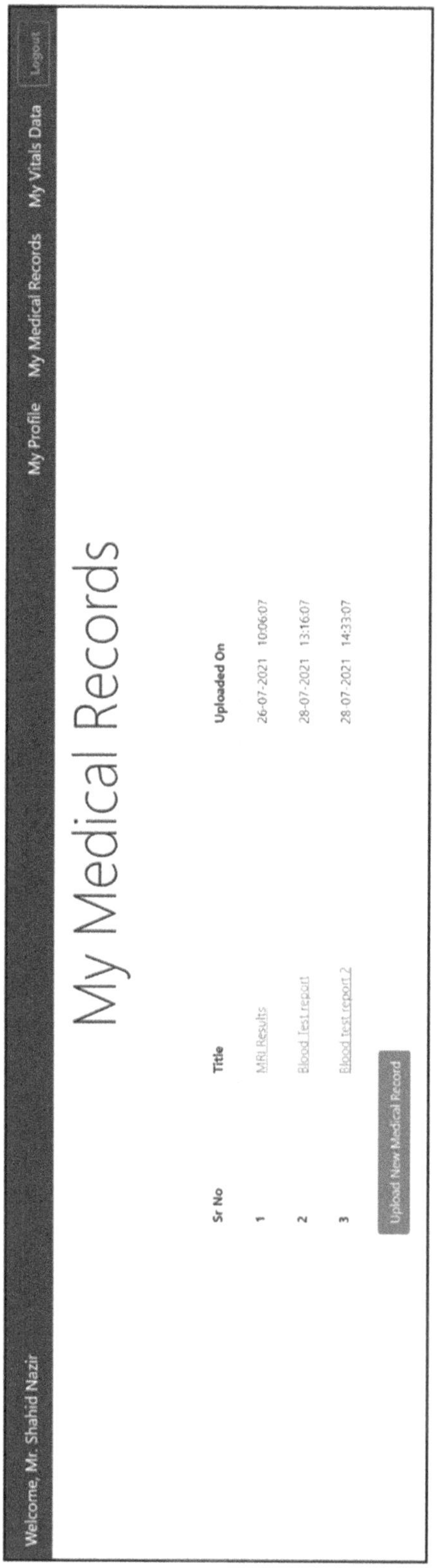

FIGURE 9.15
Patient records.

FIGURE 9.16
Medical record and access control.

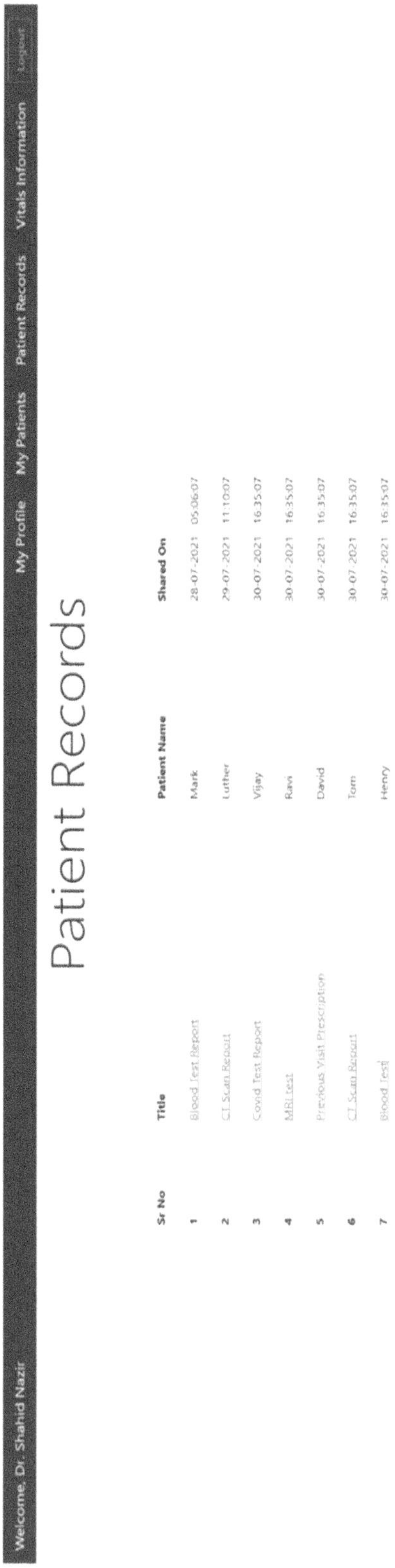

FIGURE 9.17
Doctor's view of patient records.

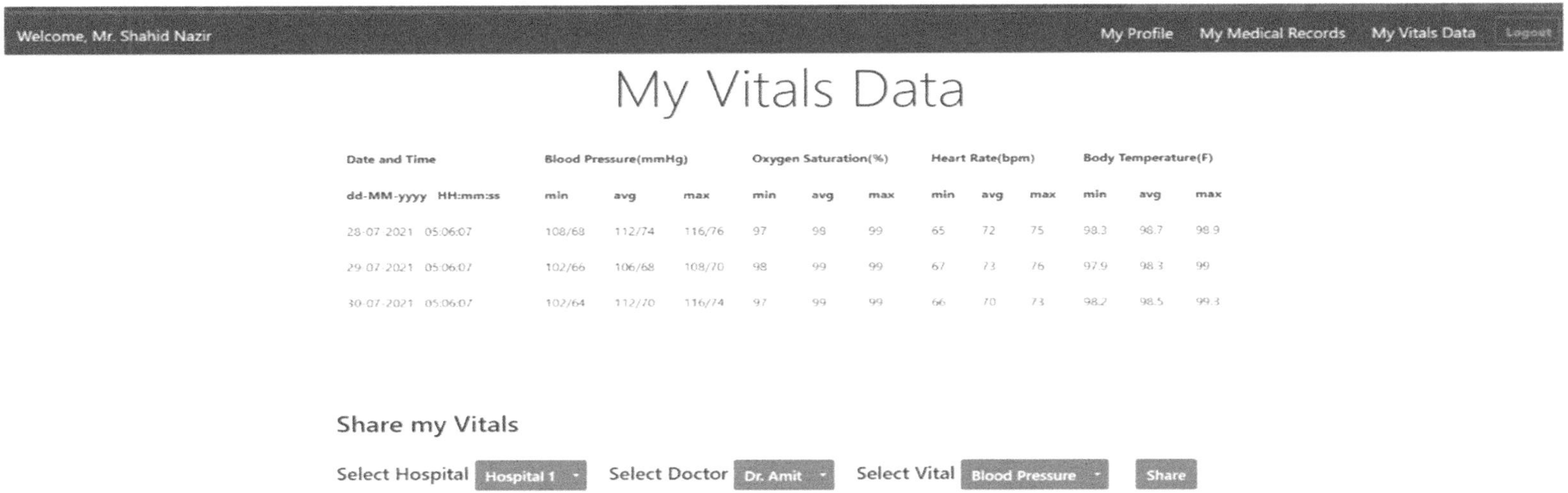

Welcome, Mr. Shahid Nazir | My Profile | My Medical Records | My Vitals Data | Logout

My Vitals Data

Date and Time	Blood Pressure(mmHg)			Oxygen Saturation(%)			Heart Rate(bpm)			Body Temperature(F)		
dd-MM-yyyy HH:mm:ss	min	avg	max	min	avg	max	min	avg	max	min	avg	max
28-07-2021 05:06:07	108/68	112/74	116/76	97	99	99	65	72	75	98.3	98.7	98.9
29-07-2021 05:06:07	102/66	106/68	108/70	98	99	99	67	73	76	97.9	98.3	99
30-07-2021 05:06:07	102/64	112/70	116/74	97	99	99	66	70	73	98.7	98.5	99.3

Share my Vitals

Select Hospital [Hospital 1] Select Doctor [Dr. Amit] Select Vital [Blood Pressure] [Share]

FIGURE 9.18
Patient's vitals.

References

[1] What is Blockchain Technology? – IBM Blockchain. (n.d.). Retrieved June 17, 2021, from https://www.ibm.com/in-en/topics/what-is-blockchain

[2] Androulaki, E., A. Barger, V. Bortnikov and J. Yellick (2018) "Hyperledger Fabric: A Distributed Operating System for Permissioned Blockchains", Unknown. https://www.researchgate.net/publication/322851452_Hyperledger_Fabric_A_Distributed_Operating_System_for_Permissioned_Blockchains

[3] Hyperledger Fabric – Hyperledger. (n.d.). Hyperledger. Retrieved June 17, 2021, from https://www.hyperledger.org/use/fabric

[4] Mohanta, B. K., D. Jena, S. S. Panda and S. Sobhanayak, "Blockchain technology: A Survey on Applications and Security Privacy Challenges," *Internet of Things* 8 (2019): 100107. 10.1016/j.iot.2019.100107

[5] Pap, I. A., S. Oniga, I. Orha and A. Alexan, "IoT-based eHealth Data Acquisition System," 2018 IEEE International Conference on Automation, Quality and Testing, Robotic AQTR 2018 - THETA, 2018, pp. 1–5. https://ieeexplore.ieee.org/document/8402711

[6] Yew, H. T., M. F. Ng, S. Z. Ping, S. K. Chung, A. Chekima and J. A. Dargham, "IoT Based Real-Time Remote Patient Monitoring System," 2020 16th IEEE International Colloquium on Signal Processing & Its Applications (CSPA), 2020, pp. 176–179. https://ieeexplore.ieee.org/document/9068699

[7] Huh, S., S. Cho and S. Kim, "Managing IoT Devices using Blockchain Platform," 2017 19th International Conference on Advanced Communication Technology (ICACT), 2017, pp. 464–467. https://ieeexplore.ieee.org/document/7890132

[8] Dubovitskaya, A., F. Baig, Z. Xu, R. Shukla, P. S. Zambani, A. Swaminathan, M. M. Jahangir, K. Chowdhry, R. Lachhani, N. Idnani, M. Schumacher, K. Aberer, S. D. Stoller, S. Ryu and F. Wang (2020), "ACTION-EHR: Patient-Centric Blockchain-Based Electronic Health Record Data Management for Cancer Care", https://www.jmir.org/2020/8/e13598/

[9] Henrique, A., M. C. A. da Costa and R. Da R. Righi (2019). "Electronic Health Records in a Blockchain: A Systematic Review", https://www.researchgate.net/publication/336162008_Electronic_health_records_in_a_Blockchain_A_systematic_review/

[10] Shahnaz, A., U. Qamar and A. Khalid (2019). "Using Blockchain for Electronic Health Records", https://ieeexplore.ieee.org/abstract/document/8863359/

[11] Zhang, P., J. White, D. C. Schmidt, G. Lenz and S. T. Rosenbloom, "FHIRChain: Applying Blockchain to Securely and Scalably Share Clinical Data." *Computational and Structural Biotechnology Journal* 16 (Jul. 2018): 267–278. https://www.sciencedirect.com/science/article/pii/S2001037018300370/

[12] Kuo, T.-T., H.-E. Kim and L. Ohno-Machado, "Blockchain Distributed Ledger Technologies for Biomedical and Health Care Applications." *Journal of the American Medical Informatics Association* 24 (2017) (6): 1211–1220. https://www.researchgate.net/publication/321147639_Blockchain_distributed_ledger_technologies_for_biomedical_and_health_care_applications

[13] Pirtle, C., and J. Ehrenfeld, "Blockchain for Healthcare: The Next Generation of Medical Records?" *Journal of Medical Systems* 42 (Sep. 2018) (9): 172. https://www.researchgate.net/publication/326959042_Blockchain_for_Healthcare_The_Next_Generation_of_Medical_Records

[14] Sun, J., L. Ren, S. Wang and X. Yao, (2020) "A Blockchain-based Framework for Electronic Medical Records Sharing with Fine-Grained Access Control". https://journals.plos.org/plosone/article?id=10.1371/journal.pone.0239946

[15] Fernández-Caramés, T. M. and P. Fraga-Lamas, (2018) "A review on the use of blockchain for the internet of things".

[16] Xia, Q., E. Sifah, A. Smahi and S. Amofa, (2017) "BBDS: Blockchain-Based Data Sharing for Electronic Medical Records in Cloud Environments". https://www.researchgate.net/publication/316354308_BBDS_Blockchain-Based_Data_Sharing_for_Electronic_Medical_Records_in_Cloud_Environments

[17] Nguyen, D. C., P. N. Pathirana, M. Ding, Seneviratne, (2019), "A. Blockchain for Secure EHRs Sharing of Mobile Cloud Based E-Health Systems". https://ieeexplore.ieee.org/document/8717579

[18] Zhang, P., J. White, D. C. Schmidt, L. Gunther and S. T. Rosenbloom, FHIRChain: "Applying Blockchain to Securely and Scalably Share Clinical Data". vol. 16, 2018, pp. 267–278. https://www.sciencedirect.com/science/article/pii/S2001037018300370

[19] Jeong, S., J.-H. Shen and B. Ahn (2021), "A Study on Smart Healthcare Monitoring Using IoT Based on Blockchain". https://www.hindawi.com/journals/wcmc/2021/9932091/

[20] Srivastava, G., J. Crichigno and S. Dhar (2019), "A Light and Secure Healthcare Blockchain for IoT Medical Devices". https://www.ieeexplore.ieee.org/document/8861593

10

Applications of Blockchain Technology to Emerging Educational Scenario: International Context and Indian Potentialities

P. K. Paul

Executive Director, MCIS, & Assistant Professor
(IST), Department of CIS, & Information Scientist
(Offg.), Raiganj University, India

CONTENTS

10.1 Introduction

Blockchain technology is a kind of encrypted record of dara or distributed database that deals with the data related to the transaction and contract. Blockchain is a kind of distributed database that needs an independent record. In blockchain technology, digital ledgers play an important role and here such digital ledger is accessible in different platforms, and not bound to be in a particular place [1–3]. Blockchain technology has different kinds of financial activities and services, and it is directly and indirectly helps in promoting digital currency viz. bitcoin transactions. The tangible and intangible assets can be tracked and recorded within the network and ledger of blockchain. This is a worthy tool, technique, and procedure in respect of financial management and business transaction management. Blockchain technology in the recent past become a field of study and gaining popularity as an academic program internationally. In blockchain technology, the financial transaction become possible effectively without any help in third-party affiliations or need. Since there is no third party here in using blockchain technology, therefore it keeps the data encrypted; and moreover here participants have no need to share any personal data and therefore all the data is encrypted. One of the important benefits of such technology is a data breach, i.e. it significantly reduces the data breach. However, in blockchain technology, there are multiple numbers of shared copies of the same database; therefore, it helps in challenging to wage a data breach attack i.e. simply the cyber attack. Due to the fraud resistant feature, the emerging blockchain technology is holding the potential to revolutionize business and financial sector more smarter, advanced, and transparent. According to the experts, this technology is more efficient compared to the traditional business process. The rising applications and role of blockchain in different sectors, areas, and zones results in the manpower requirement in the field and as a result different educational institutions have started offering educational, training, and research programs in the areas of blockchain or allied areas [4–6]. (Figure 10.1)

10.2 Objective

The present paper entitled "Blockchain Technology: Applications to Emerging Educational Scenario—*International Context and Indian Potentialities*" is conceptual and policy based. The core aim and agenda of this paper are included as follows:

Basic Blockchain Features
Blockchain is a specific type of database.
It store data in blocks that are then chained together and therefore differs from traditional databases.
The new data entered into the fresh block and once full then it chained with previous block and stays chronological order
Though blockchain can be used for various data types but it is most worthy in ledger for transactions.
Blockchain is used in a decentralized therefore no single person or group has control; however all the user having chained control.
Decentralized blockchains are immutable i.e. entered data is irreversible. However in Bitcoin, transactions are permanently recorded and also can be viewed to anyone.

FIGURE 10.1
Few development phases of blockchain from concept to a blockchain technology.

- To learn about the basics of the blockchain technology including its foundation and historical background.
- To know about the features, characteristics, and basic functions of the blockchain technology in brief.
- To dig out the basic applications of the blockchain technology in a contemporary world in a changing scenario.
- To learn about the emerging application of blockchain technology including allied technologies.
- To know about the blockchain technology related to an educational program in an international scenario with reference to India.
- To find out the potential academic and research program in the areas of blockchain technology in an Indian context.
- To learn the main issues and challenges in blockchain technology in an international and Indian context.

10.3 Methodology Adopted

The present paper is theoretical and conceptual in nature and it is aimed to learn about the basics of the blockchain technology including secondary sources and primary sources. Secondary sources are studied to know about the basics of the blockchain technology including features, functions, and roles. The aspects such as current scenario of educational programs, academic, and research degree are availed from the web survey and review. The search engine Google is used to learn about the academic program of blockchain technology; and in this context five search result pages considered important in respect to the search query "MSc/Masters in Blockchain Technology," "MTech in Blockchain Technology," and "BSc/Bachelors in Blockchain Technology." For proposed programs in an Indian context, the website and regulations of the regularity and controlling bodies (viz. AICTE, UGC) have been studied.

10.4 Blockchain Technology: Foundation to Basic Applications and Emergence

Blockchain technology is becoming a technology and has reached from the basic term *blockchain*. As far as the scientific term is concerned, blockchain is not only a database; furthermore, it is a new technology with more digital trust [7,8]. In the year 1982, David Cham coined and proposed the term *blockchain* and later Stuart Haber and W. Scott Stornetta, in the year 1992 wrote a book on consortium where more concepts and ideas became introduced on blockchain [6,9,10]. However, first blockchain network was invented and implemented by Satoshi Nakamoto after deploying the first digital currency, Bitcoin. It is important to note that blockchain technology is the technology behind blockchain, and it cannot be owned. And further development phases are depicted in Figure 10.2.

Some of the Significant Development in respect of Blockchain Technology	
Year	**Development Phase**
1991	A cryptographically secured chain of blocks is described for the first time by Stuart Haber and W Scott Stornetta
1998	Computer scientist Nick Szabo works on 'bit gold', a decentralised digital currency
2000	Stefan Konst publishes his theory of cryptographic secured chains, plus ideas for implementation
2008	Developer(s) working under the pseudonym Satoshi Nakamoto release a white paper establishing the model for a blockchain
2009	Nakamoto implements the first blockchain as the public ledger for transactions made using bitcoin
2014	Blockchain technology is separated from the currency and its potential for other financial, interorganisational transactions is explored. Blockchain 2.0 is born, referring to applications beyond currency

FIGURE 10.2
Few development phases of blockchain from concept to a blockchain technology.

The tools, technologies, concept, and protocols used behind the blockchain are blockchain technology. Blockchain is helpful in developing cryptocurrencies (i.e. digital currency and secured by the cryptography) similar to the Bitcoin. The blockchain is not possible to change i.e. immutable, here transaction and data stored in multiple places on a computer network and the strategies used is beyond the cryptocurrencies [11–15]. The blockchain can be three types viz. public blockchain, private blockchain, and hybrid blockchain.

Public blockchain is a kind of blockchain that is open in nature and means here decentralized computer networks are used for accessibility to anyone interested in a transaction [16]. Those who are validated in the transaction receive rewards. Such kinds of blockchain are used in two models:

- Proof-of-work
- Proof-of-stake

Furthermore, two common examples of public blockchains include the Bitcoin and Ethereum (ETH) blockchains. High security, open environment, anonymous nature, no regulations, full transparency, distributed, etc. are the prime examples of blockchain technology [17], [18].

Whereas *private blockchains* have the restrictions and are not open, they have access restrictions. In private blockchains, people need permission to do the transaction and here the role of system administrator is important. Hyperledger is a private, permissioned blockchain. Basically, private blockchain solutions develop these platforms for the internal networking system of a company. Some of the features of private blockchain technology are full of privacy, high efficiency, faster transaction, better scalability, faster, and speediness [8,19,20].

The hybrid blockchains are the mix of both public blockchain and private blockchain. Hybrid blockchains offer better control for achieving a higher goal and contain centralized and decentralized features. It is important to note that, though hybrid blockchains are not open but still offer blockchain features such as integrity, transparency, and security. A hybrid blockchain allows maximum customization and is select data or records from the blockchain can be allowed to go public, keeping the rest as confidential in the private network [21–23].

Blockchain technology is useful in diverse areas and sectors and among these few important areas are financial services, health care, government, travel and hospitality, retail, etc. Blockchain technology is applicable in diverse areas due to its role and uses:

- In secure and proper sharing of medical and health data
- In the areas of NFT marketplaces
- Regarding the music royalties tracking
- In the areas of cross-border payments
- In proper and real-time IoT operating systems
- In enhancement of the personal identity security
- In regard to the anti-money laundering tracking system
- In proper and sophisticated supply chain and logistics monitoring
- In proper logistics monitoring
- Regarding the voting mechanism
- In advertising and allied areas
- Regarding original content and information creation
- Regarding cryptocurrency exchange
- In real estate processing platform and systems [24,25]

However, blockchain technology is emerging and growing rapidly in the following diverse areas most significantly.

10.4.1 In Financial Sectors and Services

The rising applications of the blockchain technology in different financial sectors is already planned and started to get practice in different areas of financial sectors in various innovative ways [26,27]. Blockchain technology simplifies the whole process associated with asset management and payments by providing an automated trade life cycle. In regard to uses of blockchain technology, all the participants would have access to the exact same data about the transaction is important. The applications of this technology reduced the requirement of the middleman or broker in respect to financial management in blockchain technology in the areas of financial asset management, transaction management, etc.

As far as the banking sector is concerned, blockchains are actively engaged in high-tech securities instead of traditional methods of securities. The banking industry has started experimenting in banking and similar sectors on blockchain applications in different contexts. Blockchain is an open-source software that helps in support of digital assent and financial transfer in real-time scenarios. Therefore, the biggest advantages of blockchain technology in this sector include efficiency, healthy security, unchangeable records, and quick and proper transaction time. And mainly less third-party involvement.

10.4.2 In Healthcare Services and Sectors

Blockchain in the healthcare and medical segment brings several benefits such as privacy, security, and interoperability and further it is dedicated in many interoperability challenges in health care. Blockchain technology helps in secure sharing of healthcare data from different entities and people involved in the process. It, furthermore, reduces or eliminates the interference of a third party including additional costs in health care. Using blockchain technology healthcare-related data and records can be stored with proper encryption in a distributed way. Furthermore, the implementation of the digital signatures also ensures the privacy and authenticity [26,28,29]. As a whole, blockchain technology in health care and medical systems brings the following:

- Managing and disseminating electronic and digital medical record or data
- Protection and secure of healthcare and medical data
- Personal (admin and staff) and patient health record data management
- In proper point-of-care genomics management
- In proper electronics, health records and data management

10.4.3 In Sophisticated Governmental Activities

Blockchain technology is important in the field of government and administration, including the transformation of various operations and services. Blockchain technology in the government sector helps in improving the data transactional challenges and enables better management of data between multiple organizations, branches, and departments. The transparency and better monitoring and audit of the transactions become effective with blockchain technology. According to the experts, following are three major uses of blockchain technology in the government section.

10.4.3.1 In Building Trust

As far as secure sharing of data between citizens and agencies is concerned, blockchain technology is helpful in reducing any such concerns and ultimately such an initiative can impact building trust [30,31].

10.4.3.2 In Reducing Data Leaks

In reducing the time, cost, and risk of sensitive data management, blockchain technology is very important. With blockchain technology, it is able to provide an immutable and transparent audit trail and ultimately such are worthy in regulatory compliance. Moreover, it helps in contract management, identity management, and improving citizen services.

10.4.3.3 In Removing Barriers to Innovation

Blockchain technology is helpful in imperfect information including restrictive regulation, institutional inertia, and cybersecurity issues. Furthermore, in disruption of new business models of government and administration, blockchain technology is worthy.

10.4.4 In Retail Management Sectors

Blockchain technology gives opportunities in retail sectors and that includes authenticity of high-value goods, preventing retail fraudulent business transactions, issuing virtual warranties, and streamlining operations related to the supply chains. Furthermore, blockchain technology helps in reliability, sustainability, as well as quality of tracking. It helps in management processes that help in healthy retail informatics. However, there are key issues in respect of blockchain technology in retail management and sectors:

- Changing consumer experience
- Poor customer loyalty maintenance
- Inefficient internal communication
- Fraud and counterfeit goods
- Poor marketing strategy
- Retail price inflation
- No uses in advanced technologies [32–34].

The scenario of blockchain technology in the retail segment including retail management and marketing management is noticeable in developing countries as well.

10.4.5 In Travel and Hospitality Sectors

The growing applications of blockchain technology in the travel and hospitality sectors are important to notice. The applications of blockchain technology in travel, tourism, and hospitality sectors are rising:

- Money transactions
- Storing some of the important documents such passports
- Different types of identification cards
- Regarding reservations and managing travel insurance including rewards, etc. [9,35]

These are the prime examples in blockchain technology applications; however, in certain other areas as well the applications have been raised: entertainment, transportation, corporate training and education, industries, etc. (as depicted at Figure 10.3).

10.5 Key Challenges of Blockchain

Blockchain technology, though popular, has some of the challenges and issues in respect of development, implementation, and progress. There is limited awareness and understanding regarding the blockchain concept and is considered a key challenges regarding the use of blockchains in a full-fledged manner in industries other than the financial services sector. The challenges regarding the legacy infrastructure include lack of proper technical understanding of blockchain in the mainstream is also an important point to note. The cultural shift from the traditional to almost scientific and technologically sound systems is the major shift of decentralizing the whole process; therefore, this can be considered as vital in this context. The regulations regarding the blockchain technology are an important issue and in this context existing regulations need to check out, including the data privacy laws [36,37]. Furthermore, the security-related issues for the shared databases also add to the major roadblocks in adopting blockchain. The international market and industry

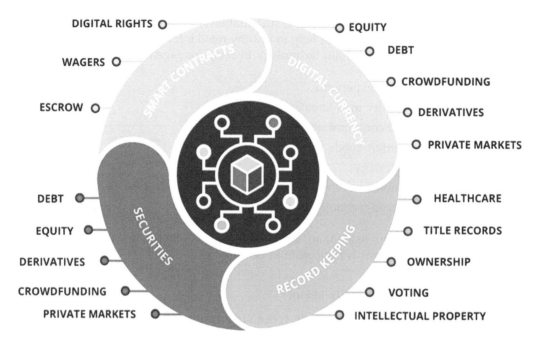

FIGURE 10.3
Some of the basic and emerging applications of blockchain technology.

world is not yet ready to implement blockchain technology in fullest manner; however, that is the need of hour [25,37].

10.5.1 Manpower and Human Resource Development: Blockchain Technology

Internationally, universities and higher educational institutions have moved into strategic development into offering the latest, emerging, interdisciplinary, skill-based subjects and degrees and as a result various emerging subjects are listed at the course/program of study with bachelors, masters, MPhil, PhD, and other professional and research-based degrees. Among the emerging and latest subjects, a few important ones are cloud computing, big data, cybersecurity, information assurance, Internet of Things (IoT), data analytics, etc. And within this list blockchain technology has become an important subject of study and available as individual blockchain technology or allied nomemclature. Some of the nomenclatures in this regard include:

- Blockchain
- Blockchain technology
- Blockchain and financial technology
- Blockchain management
- Blockchain and distributed technology
- Blockchain and distributed ledger technology
- Cryptography and blockchain technology
- IoT, cybersecurity with blockchain technology

Furthermore, as per the study (as depicted in the methodology portion), the degrees offered for such subjects are master of science (both MS & MSc), master of business administration, and master of computer applications. The programs are available as an individual study and also as a specialization at the related programs. Up to seven page results as per search term selected show that most of the blockchain technology related masters programs are offered in developed countries, while as per the methodology adopted only Amity University in India (a developing nation) offers MCA with blockchain technology specialization as far as masters is concerned (Refer Table 10.1).

As per the study, it has noted that blockchain technology is offered as a specialization in only two programs while others are fully on the subject; therefore, there is a need to offer blockchain technology and allied programs as a specializations at the allied programs. The masters programs are offered in minimum numbers and bachelors degrees reveal higher/maximum numbers as per the methodology adopted and depicted in Table 10.2.

TABLE 10.1

Some of the Academic Programs of Blockchain Technology at the Masters Level as per Methodology Adopted

Sl. No.	Degrees Offered	Universities	Remarks	Country
1	MS Information Systems (Blockchain Technology)	Northeastern University	32 Credit (2 Years), Offered at Silicon Valley Campus	USA
2	MSc Blockchain Application & Financial Technologies	University of East London	1 Year (FT) & 2 Years (PT), Open to Physical Science/Computing Bachelors	UK
3	MSc Blockchain	Dublin City University	2 Year (PT), Online Mode, Open to Computer related Bachelors	UK
4	MSc Blockchain & Distributed Ledger Technologies (Business & Finance)	The University of Malta	3 Years (FT), Oncampus, Open to Computing & Business related Bachelors	Malta
5	Master of Blockchain Enabled Business	RMIT University	1.5 Year (FT) & 3 Years (PT), Oncampus & Online, Open to Business related Bachelors	Australia
7	Master in Global Business with Blockchain	The University of the Cumberlands	31 Credit, Fully Online, Bachelors in Any Stream	USA
8	Masters in Blockchain Technologies	The Polytechnic University of Catalonia	1 Year, Blended, 60 Credit	Spain
9	Master in Blockchain Technologies	Global Institute of Leadership & Technology (The University of Barcelona)	1 Year, Fully Online	Spain
10	MBA Blockchain Management	European University Business School	1 Year, 90 Credit	Switzerland, Spain & Germany
11	MCA Blockchain	Amity University	2 Year, Computing or any Bachelor holder	India

TABLE 10.2

Some of the Academic Programs on Blockchain Technology at the Bachelors Level as per Methodology Adopted

Sl. No.	Degrees Offered	Universities	Remarks	Country
1	BSc IT (Blockchain Technology)	Techno India University	3 Years (FT), Any 10+2 Candidate	India
2	BSc IT (Blockchain Technology)	Parul University	3 Years (FT), Any 10+2 Candidate	India
3	BSc IT (Blockchain Technology)	Maulana Abul Kalam Azad University of Technology	3 Years (FT), Any 10+2 Candidate	India
4	BSc CS (Cryptography &Blockchain)	Karnavati University	3 Years (FT), Science 10+2 Candidate	India
5	BTech CSE (Cyber Security & Blockchain Technology)	SASTRA Deemed University	4 Years (FT), Science 10+2 Candidate	India
7	BTech CSE (Cyber Security & Blockchain Technology)	Ajeenkya DY Patil University	4 Years (FT), Science 10+2 Candidate	India
8	BTech CSE (Blockchain & Distributed Computing)	Vivekananda Global University	4 Years (FT), Science 10+2 Candidate	India
9	B.Engg. CSE (Blockchain Technology)	Chitkara University	4 Years (FT), Science 10+2 Candidate	India
10	B.Tech. CSE (Blockchain Technology)	University of Petroleum and Energy Studies	4 Years (FT), Science 10+2 Candidate	India
11	B.Tech. CS (Blockchain Technology)	Sharda University	4 Years (FT), Science 10+2 Candidate	India
12	B.Tech. CSE (Blockchain Technology)	SRM Deemed University	4 Years (FT), Science 10+2 Candidate	India
13	B.Tech- Blockchain Technology	Swarnim Startup & Innovation University	4 Years (FT), Science 10+2 Candidate	India
14	B.Tech-CSE (Blockchain Technology)	RIMT University	4 Years (FT), Science 10+2 Candidate	India
15	B.Tech-CSE (IoT,Cyber Security with Blockchain Technology)	Manakula Vinayagar Institute of Technology	4 Years (FT), Science 10+2 Candidate	India
16	B.Tech-CSE (IoT,Cyber Security with Blockchain Technology)	PSCMR College of Engineering & Technology	4 Years (FT), Science 10+2 Candidate	India
17	B.Tech-CSE (Blockchain Technology)	The Northcap University	4 Years (FT), Science 10+2 Candidate	India
18	BTech CS (Blockchain & Distributed Computing)	Srinivas University	4 Years (FT), Any 10+2 Candidate	India

TABLE 10.2 (Continued)

Some of the Academic Programs on Blockchain Technology at the Bachelors Level as per Methodology Adopted

Sl. No.	Degrees Offered	Universities	Remarks	Country
19	B.Tech-CSE (Blockchain Technology)	Chandigarh University	4 Years (FT), Science 10+2 Candidate	India
20	B.Tech-CSE (IoT,Cyber Security with Blockchain Technology)	Vasireddy Venkatadri Institute of Technology	4 Years (FT), Science 10+2 Candidate	India
21	B.Tech-CSE (IoT,Cyber Security with Blockchain Technology)	Dronacharya Institute of Technology	4 Years (FT), Science 10+2 Candidate	India
22	B.Tech-CSE (IoT,Cyber Security with Blockchain Technology)	Gulzar Institute of Technology	4 Years (FT), Science 10+2 Candidate	India
23	B.Tech-CSE (Blockchain Technology)	Adamas University	4 Years (FT), Science 10+2 Candidate	India
24	B.Tech-CSE (Blockchain Technology)	Poornima University	4 Years (FT), Science 10+2 Candidate	India
25	B.Tech-CSE (IoT,Cyber Security with Blockchain Technology)	Sri Ramchandra Institute of Higher Education & Research Deemed University	4 Years (FT), Science 10+2 Candidate	India

An interesting fact revealed from the search keywords of BSc/Bachelors in blockchain technology was that almost all seven pages of Google results belongs to the institutions offering blockchain technology from the country of India.

The study shows that blockchain technology and allied fields are emerging in India with mostly B.Tech. (Bachelor of Technology) degree and expect Chitkara University offers bachelor of engineering nomenclature instead of bachelor of technology. Another important fact noted according to the study is that most of the blockchain technology of bachelors comes as a specialization with computer science and engineering subjects. And most of such degrees are four years' duration with eligibility as 10+2 with pure science, except Srinivas University, Mangalore, India, which offers a program to any 10+2. BSc programs shown are three years.

10.6 Carrier, Curriculum, and Employability in Respect of Blockchain Technology

Blockchain technology is rising popularly around the world and as a result the program became popular in all the industries and business sectors along with other emerging technologies such as cloud computing, big data, cybersecurity, information assurance, Internet of Things (IoT), data analytics, etc. Blockchain and financial technology is of major importance to the government and businesses in all sectors and, due to shortage of the manpower, various programs started internationally that can be connected with the technology industry, related banking and financial sector, and other sector specific areas such as agriculture, health care, manufacturing, and automotive. Apart from the technology space and business world, the blockchain technology educated can lead to research jobs and areas.

- Blockchain Developer
- Blockchain Architect
- Blockchain Engineer
- Blockchain Analyst
- Blockchain UX/UI Designer
- Blockchain Product Manager
- Charted Blockchain Expert
- Charted Digital Asset Analyst
- Business Intelligence/Business Analyst
- Senior Software Architect
- IT Consultant/IT Program Manager
- System Developer
- Cyber Security Specialist/IT Security Consultant
- IT Infrastructure Manager
- Cloud Architect
- Blockchain Business Analyst
- Financial Analyst
- Crypto Community Manager
- Blockchain Researcher
- Distributed Computing – Algorithms Developer
- Distributed computing – Technical Architect
- Distributed computing – Data Engineer
- Blockchain Consultant
- Blockchain Developer
- Blockchain Solution Architect
- Cryptocurrency Analyst
- Research Analyst: Blockchain
- Cryptocurrency Developer
- Cryptocurrency Mining Engineer
- Cloud Engineer with Bitcoin Protocol/Blockchain
- Bitcoin Full-Stack Developer

Apart from the banking, global finance, supply chain management, health care, digital media, and beyond the blockchain professionals can help in getting jobs in research industries as well. As per LinkedIn's annual Emerging Jobs 2018 report, blockchain developer is one of the most coveted jobs among all occupations. Earning a blockchain degree leads the way forward in this exciting field—and unlock career opportunities as you go. The blockchain technology graduates need to develop their self in such a way so that they can be able to perform the following:

- Developing programming code in crypto assets as well as in performing blockchain data mining
- Create and implementation of the cryptocurrencies including hyper ledgers, decentralized applications
- Able in analyzing the blockchain technology's role, and implementation in the areas of business, society, and industries.
- To perform the specific needs and challenges of the blockchain technology and allied technologies as per current trends.
- To be able to solve management and allied problems in organizations and institutions.

Therefore, blockchain technology programs at bachelors, masters, and doctoral levels need proper attention in designing and developing curriculum as per the need and their role and job descriptions. The degree holders should gain knowledge regarding information technology, management, mathematics and statistics, and business and societies. As per the market demand and as per the trend of blockchain in the industries and organizations, some of the programs need to start in Indian context at par global trend. Table 10.3 depicts the proposed program on blockchain with ME/MTech.

Such M.Tech./ ME-Bblockchain technology may be considered as worthy for those who is to be completed B.Tech.-blockchain technology (since BTech in the field already exists) or B.Tech. for any appropriate branches. However, Table 10.4 depicts some of the blockchain programs with CS/computing subjects as specializations.

TABLE 10.3

Proposed Programs of Blockchain with ME/MTech

Proposed Blockchain Technology Program Potentiality with MTech/ME Program Level
MTech/ME- Blockchain Technology
MTech/ME- Blockchain and Financial Technology
MTech/ME- Business Informatics with Blockchain Technology
MTech/ME- Blockchain Technology & IoT

TABLE 10.4

Proposed Programs of Blockchain with CS/Computing Subjects

Proposed Blockchain Technology Program with Computing/Computer Science at Masters Level
MSc-Computer Science (Blockchain Technology)
MSc-Computing by Research (Blockchain Technology)
MSc-Computer Science by Research (Blockchain Technology)
MSc-Computing by research (Blockchain Technology)

TABLE 10.5

Proposed Programs of Blockchain with IT/IS/IST Subjects

Proposed Blockchain Technology Program with Information Technology/Information Science
MSc-Information Technology (Blockchain Technology)
MSc-Information Science (Blockchain Technology)
MSc-Information Science & Technology (Blockchain Technology)
MSc-Computer & Information Science (Blockchain Technology)

TABLE 10.6

Proposed Programs of Blockchain at the Research Level

Proposed Blockchain Technology Program at Research Level/Degrees
PhD-Blockchain Technology
PhD-Blockchain and Distributed Technology
PhD-Computer Applications (Blockchain Technology)
PhD-CSE (Blockchain Technology)
PhD-IT/IS (Blockchain Technology)

However, Table 10.5 depicts some of the possible programs in the subject of information technology, information science, and information science and technology context in application context rather than blockchain systems development context, which might be better with computer science/CSE context.

Blockchain technology is emerging internationally; therefore, it is worthy and beneficial in research context and therefore programs at the research level need to start for institutions and countries, if not yet. Some of the proposed programs are listed in Table 10.6.

Blockchain technology is rising gradually in all the sectors and areas and it is helpful in leading our society into knowledge society and knowledge economy with full of digital support. The proposed programs are depicted at the current trends and nomenclature as already offered by the governing bodies and authorities. In the future, programs related to the blockchain technology are also required at the management- and commerce-related subjects.

10.7 Conclusion

Blockchain technology is gaining popularity around the world due to its importance and significance. As blockchain is a peer-to-peer decentralized distributed ledger technology, therefore it is worthy in making digital asset transparent and furthermore unchangeable. As it works without any third party or intermediary, therefore the concept of brokers is decreasing. It is an emerging technology and thus it is attracting a lot of public attention to the financial market and organizations due to the reduced risks and frauds in a scalable manner. Many countries are putting their efforts in popularization of blockchain technology in wider markets and organizations. Educational institutions, research centers, business and commercial associations, and merchants are also using efforts in developing

Section C

Security Aspects in Blockchain and IoT

11

Cryptographic and Consensus Techniques Supporting Privacy and Security Management of Cryptocurrency Transactions

Tun Myat Aung[1] and Ni Ni Hla[2]

[1]*University of Information Technology – UIT, Yangon, Myanmar*
[2]*University of Computer Studies, Yangon – UCSY, Myanmar*

CONTENTS

DOI: 10.1201/9781003203957-14

11.1 Introduction

A cryptocurrency, also known as digital currency, is a digital financial asset considered to serve as a type of exchange medium in which individual records of coin ownership are kept in a ledger, that is a kind of distributed electronic database which utilizes powerful cryptographic techniques to make transaction records secure, manage the formation of new coins, and verify the transfer of coin ownership. It is usually not available in tangible form, such as money notes, and it is not dispensed by a central authority. Cryptocurrencies, in contrast to centralized digital money and central banking systems, generally rely on decentralized power. Every cryptocurrency is driven by a blockchain, a distributed ledger technology that acts as a database of public financial transactions.

A blockchain is a recent technology that integrates data structure techniques, cryptographic techniques, distributed database techniques, consensus techniques, and a peer-to-peer (P2P) network. A blockchain may be considered a distributed ledger that is open, secure, and reliable because of these powerful techniques. Furthermore, these approaches enable the ongoing connecting of transactions to blockchain, which creates a jointly maintained ledger that keeps track of all transactions and historical data. Internet users who have never met can create a credit arrangement using a P2P ledger with no central authority.

Bitcoin is the most popular cryptocurrency, which is a system for electronic payments that is distributed. It was designed by Satoshi Nakamoto [1]. Since the birth of Bitcoin, a growing number of retailers have stated their readiness to allow it as a payment method. It is possible to save, transmit, and trade bitcoins from one Bitcoin network participant to another. The transactions can be carried out without the need of middlemen. Bitcoins can also be traded without the assistance of banks or clearing houses. As a result, Bitcoin has had a huge economic and technological impact throughout the world. There are many kinds of cryptocurrencies known as cryptotokens, which are being traded in the CoinMarketCap [2]. "The most significant cryptotokens other than Bitcoin (BTC) are Ethereum (ETH), Litecoin (LTC), Cardano (ADA), Polkadot (DOT), Bitcoin Cash (BCH), Stellar (XLM), Chainlink (LINK), Binance Coin (BNB), Tether (USDT) and Monero (XMR)" [3]. These cryptotokens are generated by their own blockchains.

This chapter is intended for system developers and software engineers who want to gain a better understanding of the privacy and security strategies used in cryptocurrency blockchain systems. As many kinds of cryptocurrencies become more widely used today all over the world, system developers and software engineers must often analyze the security of these systems and attempt better security.

11.2 Blockchain

11.2.1 Introduction

The technology of blockchain is mainly developed for transacting a cryptotoken. Attempting to solve the problems for tracking and validation of distributed transactions by using cryptographic primitives is one of the main functions of the blockchain. The whole transaction history is recorded. If the user needs to see a recent past, they can filter it out. Blockchain has its own rules and regulations. The rules and regulations are written as a smart contract before starting a transaction. A transaction is only possible between two peers on the P2P network that operates on a cryptographic protocol. The sender side should enter a digital signature before making a transaction. This digital signature validates the transaction. If this signature is validated, the transaction is proceeded meaning that a cryptotoken should be transacted. A blockchain can transfer a digital currency directly between two parties without any third party. For use as a distributed ledger, a P2P network [4] and a distributed timestamping server [5] are used to manage a blockchain autonomously, meaning that the whole network is decentralized. This makes blockchains suitable for the recording of transactions of any kind. As a result, the blockchain provides a number of advantages to a cryptotoken business, including "immutability, decentralization, anonymity, improved security, and higher capacity" [6].

11.2.2 Structure of a Block

The blockchain is a data structure that represents a public distributed ledger made of a linked list of blocks that are connected together using cryptographic primitives. A fundamental unit of blockchain is the block, which is made of two parts: "the block header and the block body. The block header consists of block version, previous block hash, Nonce, Merkle root hash, and timestamp. The block body contains a transaction counter and transactions" [7]. The structure of a block in a blockchain is depicted in Figure 11.1. For the formation of blockchain, the blocks are joined as a linked list by the hash value of the previous block header, as seen in Figure 11.2, and are cryptographically protected.

11.2.3 Blockchain Architecture

Blockchain architecture is made of "six layers: a data layer, a network layer, a consensus layer, an incentive layer, a contract layer, and an application layer" [8–10], which are depicted in Figure 11.3.

- *Data Layer* – At the data layer, transaction data is collected and stored in a block of the linked list data structure known as a ledger utilizing asymmetric encryption methods, hash functions, Nonce and time-stamped Merkle trees [8,9].
- *Network Layer* – The network layer connects all of the nodes in the network and establishes a decentralized P2P network environment. It utilizes distributed systems, data transmission protocols, and data verification mechanisms. The principal responsibility of the network layer is to verify and forward the transactions over the network. A digital signature technique is utilized to verify the transaction [8,9].

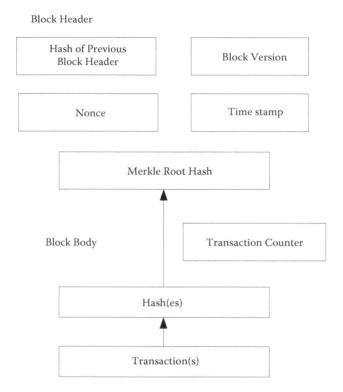

FIGURE 11.1
The structure of a block.

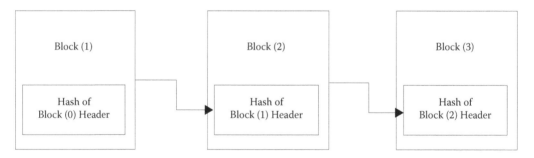

FIGURE 11.2
A sample of blockchain.

- *Consensus Layer* – Because a blockchain is a decentralized network, there is no trusted third party to verify a node's authenticity. To solve this, a consensus method is utilized for decentralized nodes. The consensus layer ensures network nodes consensus and transaction records consistency among the network nodes by using a variety of consensus protocols such as "proof-of-work (PoW), proof-of-stake (PoS), proof-of-authority (PoA)," etc. [8–10].

- *Incentive Layer* – When a miner completes the verification process and adds a block to a chain, he or she will be rewarded for their efforts. Miners will be

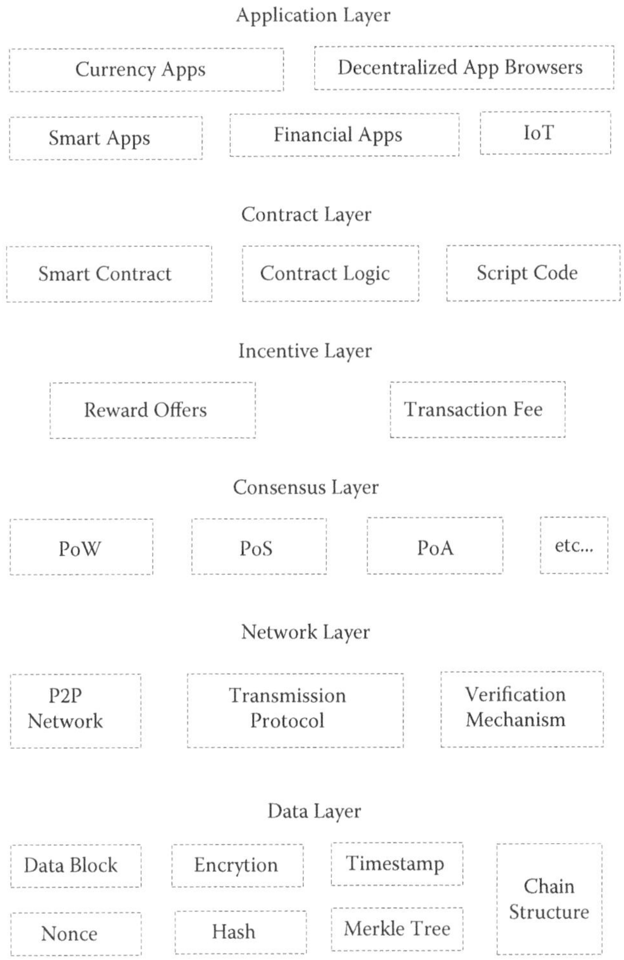

Application Layer

| Currency Apps | | Decentralized App Browsers |

| Smart Apps | Financial Apps | IoT |

Contract Layer

| Smart Contract | Contract Logic | Script Code |

Incentive Layer

| Reward Offers | | Transaction Fee |

Consensus Layer

| PoW | PoS | PoA | etc... |

Network Layer

| P2P Network | Transmission Protocol | Verification Mechanism |

Data Layer

| Data Block | Encrytion | Timestamp | Chain Structure |
| Nonce | Hash | Merkle Tree | |

FIGURE 11.3
Layer structure of a blockchain.

rewarded with incentives like digital money based on their contributions to the validation process. The incentive layer has features that specify what kind of rewards the network offers, when and how nodes can receive incentives, and the minimum transaction fees required to perform functions on the blockchain. In essence, the incentive layer responds the challenge of crowd-sourcing among distributed nodes. As a result, an effective incentive mechanism is required to ensure the greatest benefit of individuals, so that the safety of the overall blockchain system can be achieved [8–10].

- *Contract Layer* – The contract layer comprises smart contracts for managing data such as money or asset records, etc. Smart contracts reduce the number of middlemen. The contract code is kept in script file of the transaction, and after the transaction is published to the block, the code will continue to execute permanently. The contract logic will be verified by all blockchain verification nodes, and the execution outcome will be written to the blockchain [8–10].

- *Application Layer* – The application layer is made of the user interface, scripts, and APIs that serve as an interface between a client and the blockchain network. The application layer contains a variety of software applications such as smart applications, IoT applications, currency applications, financial applications, business applications, and market security applications [8–10].

11.2.4 Transaction Process

In a blockchain network, if a client wishes to transmit a cryptotoken to another client, the sender must first request a transaction. After a request has been submitted, a block representing the transaction is formed. A timestamp, hash value, nonce, block version, and data are all included in this block. The block is then submitted to all other nodes in the network. The block and the transaction are validated by every node in the network. Nodes receive rewards for the proof of work and the block is added to the chain when it has been validated. Then, the cryptotoken transfer from a client to another becomes finalized and legitimate [10–12]. The transaction process flow of a cryptotoken is depicted in Figure 11.4.

11.2.5 Wallet and Addresses

The client must first utilize a wallet app to start using a cryptotoken. A public key and a private key are generated by the wallet app. The private key may be considered as a PIN that allows us to access our cryptotokens and authorize payments, and has to be kept a secret. The public key may be considered as an account number of a bank. The private key is used to generate the public key using mathematics. A hash function is utilized to the public key to produce a cryptotoken address. Each address has its own cryptotoken balance, and transactions are generally cryptotoken transfers between these addresses. For example, if Alice transfers 50 BTC to Bob, this transaction may be considered an instruction to subtract 50 BTC from the balance of Alice's address and add 50 BTC to the balance of Bob's address [9,13].

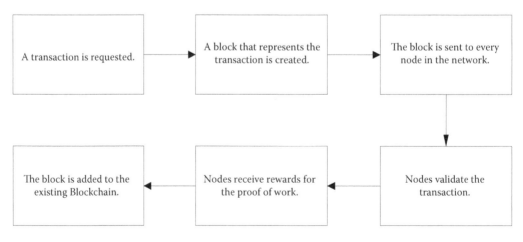

FIGURE 11.4
Transaction process flow.

11.2.6 CAP Properties

The architecture of a distributed ledger posses CAP properties: "(1) Consistency – all nodes maintain an identical ledger with the most recent update. (2) Availability – any transaction generated in the network at any point in time will be accepted in the ledger. (3) Partition tolerance – the network can continue to operate even if a branch of the nodes fails" [14]. The primary problem is that it is difficult for any commonly acceptable cryptotoken to be present without all three properties satisfied.

11.3 Privacy and Security

11.3.1 Introduction

All cryptocurrency transactions are processed on a blockchain, which is a digitized, decentralized, and public ledger. The general security features of a cryptocurrency blockchain system are determined after analyzing the requirements for privacy and security of cryptocurrency transactions.

11.3.2 Privacy and Security Requirements

The requirements for privacy and security of cryptotoken transactions are generally classified into the following eight categories.

- *Consistency of The Ledgers* – If two peers notice that a user's balance is different, that user can utilize the discrepancy to launch an attack, such as spending the same currency twice. It is subject to mistakes and inconsistencies between the ledgers of various financial institutions. Thus, the ledgers must be consistent at all times [14].
- *Integrity of Transactions* – In a blockchain, each transaction is shared across a number of users in network. Thus, the blockchain system must maintain transaction integrity and prohibit the completion of illegal transactions [14].
- *Availability of Data* – At any time and from any location, authorized users of the blockchain system should be able to access transaction data when required [14].
- *Prevention of Double-Spending* – The risk of a cryptocurrency being spent twice is known as double-spending. It occurs when a blockchain network is interrupted and cryptocurrency is effectively stolen. Because digital information may be easily reproduced by smart persons who understand the blockchain network, double-spending is a potential concern that is specific to cryptocurrency. Thus, it is essential to prevent double-spending transactions in cryptocurrency system [15].
- *Confidentiality of Transactions* – All transactions in a blockchain are made public and accessible to all participating nodes in the system. Every user on the network has the ability to start a new transaction and mine a new block. As a result, all users have access to the details of every transaction, including the sender, receiver, and amount. The more nodes are added to a network, the more transaction data would be publicly shared; and the less confidential the transaction data are. Thus, the confidentiality of transactions is essential in blockchain systems [16,17].

- *Anonymity of Users' Real Identities* – Anonymity of users' real identities are required because transactions in the cryptocurrency system are exposed to the public. The transactions are open to the public, which means that anyone can see that a certain amount is transferred from one to another. Bitcoin addresses are hashes of public keys since there is no need to use one's real identity. Thus, one's identity is the hash of his public key in Bitcoin transactions. Because these pseudo-names are used to record transactions, it is possible that the real identities aren't linked to the transactions. For complete anonymity, unlinkability must be established [18].

- *Unlinkability of Transactions* – Users wish to ensure that transactions relating to them be kept separate. When all of her transactions are linked, it is easy to gather further information about a person such as account balance, transaction type, and transaction count. Adversarial parties can accurately guess a user's real identity by combining numerical data about accounts and transactions with prior information about the person [14].

- *Difficulty Level* – To avoid fraud and assure the authenticity of transactions on a blockchain, miners check transactions and perform auditors' functions. When enough transactions have been verified by miners, a new block is appended to the blockchain. Miners might be rewarded for their work. Cryptocurrency difficulty represents the amount of computer power required to mine a block. The difficulty level of cryptocurrency determines how long it takes to find a new block. A high difficulty is required for making the blockchain network more secure against attacks [19,20].

11.3.3 Security Properties

The blockchain systems must have the following security properties to meet the privacy and security requirements given above.

- *Consistency* – Consistency refers to the feature of having the same ledger across all nodes at the same time. The blockchain, being a distributed asynchronous system, relies on a consensus technique to maintain the consistency of its shared data. When a block is legitimate in the system, all of its transactions have final consistency and cannot be changed.

- *Tamper-Resistance* – Any transaction data recorded in the blockchain cannot be altered during or after the block generation process, which is referred to as tamper-resistance.

- *Resistance to DDoS Attacks* – A denial-of-service (DOS) attack is a variant of cyber-attacks that make either a host computer or a network resource on the host unavailable to users, preventing them from accessing hosted Internet services. DoS attacks' effort is to overburden either a host system or a network resource by overflowing it with unnecessary requests, effectively shutting down legal services. A DDoS attack means "distributed" DoS attack, that is, the traffic overflowing attack coming towards a target stems from a variety of different sources distributed over the Internet. A DDoS attacker can transmit massive volumes of data to a host website by utilizing a group of hacked machines. Cryptocurrencies are a major target for DDoS attackers due to their popularity and importance [21].

- *Resistance to Double-Spending Attacks* – If inconsistencies arise as a result of duplicate cryptotoken transactions, the double-spending problem becomes a major security risk in cryptocurrency system. Thus, the cryptocurrency system must be able to resist a double-spending attack.

- *Resistance to the Majority (51%) Consensus Attacks* – The 51% attack is a technique that exploits the risks of cheatings in most consensus techniques. This attack may occur in case malicious miners exist in the cryptocurrency system. Thus, the cryptocurrency system must be able to resist a 51% attack.

- *Pseudonymity* – A condition of concealed identity is referred to as pseudonymity. Hashes of a node's (user's) public keys are used to create addresses in blockchain. Users can interact with the system anonymously by utilizing the hash values of their public keys as their pseudo-identities. Thus, the pseudonymity is considered as a privacy element to shield a user's true name [18].

- *Unlinkability* – Unlinkability is defined as the inability to specify the relationship between two observable aspects of a system with full accuracy. Deanonymization inference attacks, which connect a user's transactions together to expose the user's true identity in the presence of prior experience, are made difficult by unlinkability [14].

- *Confidentiality of Transactions and Data Privacy* – "The property that blockchain can guarantee confidentiality for all data or particular sensitive data stored on it is referred to as data privacy" [14]. The confidentiality of transaction data, including transaction content and addresses, is a popular security property common in all applications based on blockchain network. Thus, as a result of that, a blockchain and its applications include sensitive transactions and private data, privacy and confidentiality become a significant challenge.

- *Difficulty* – The difficulty refers a metric that indicates how difficult it is to mine a block in a blockchain with a hash below a certain threshold for a particular cryptocurrency. It is important to understand hash power, which reflects the combined computational power needed to mine and process the transactions on the blockchain, in order to measure the cryptocurrency difficulty of a new block [20].

11.4 Cryptographic Techniques

11.4.1 Introduction

Cryptographic techniques are used in blockchains to safeguard the identities of network users, to ensure the security and integrity of transactions, to protect all kinds of important data, to prevent from double-spending and to operate without the influence of central authorities. Hashing, asymmetric key encryptions, and digital signatures are three types of cryptographic techniques usually used in blockchains. Asymmetric key encryptions let users send and receive transaction data using a public key that everyone can see. As a result, asymmetric key encryptions are better than symmetric key encryptions in blockchains. Digital signatures are one of the major methods used to maintain the integrity and security of the information recorded in a blockchain. They are an essential element of many blockchain protocols. Hashing, the most significant feature in the blockchain, enables immutability and decentralization in a blockchain.

11.4.2 Hashing

A hash is a one-way function using mathematical stochastic transformations for mapping arbitrary-size data to a fixed-size value that is known as a hash value. A hash value is also known as a message digest. Hash functions must meet three security requirements: collision resistance, hiding, and puzzle-friendliness [12]. Generally, there are two types of hash functions: "keyed hash functions" and "unkeyed hash functions." Hash functions are principally applied to ensure data integrity [22].

A transaction list includes a lot of information about the transaction, such as the time of each transaction, the transaction number, the transaction amount, the payer, and other data. Each transaction is written and received together in the data block, allowing each transaction to be tracked back [23]. A transaction was hashed and broadcast to each node when it was completed. Thousands of transaction records might be stored in a block of each node. The Merkle tree is used by a blockchain to create the Merkle Root Hash, which is a final hash value. The Merkle tree is a binary search tree in which the tree nodes are connected together using hash references, as seen in Figure 11.5. The final hash value is stored in the block header, and it will be used to connect blocks together in a continuous chain by referencing the hash value of preceding block. If any data on the block is changed, the hash value changes, rendering all subsequent blocks invalid [6].

The hash function currently used in the most of blockchains is SHA256, which is an algorithm from a group of SHA (Secure Hash Algorithms). "The National Institute of Standards and Technology (NIST) published SHA as a Federal Information Processing Standard in the United States. The United States National Security Agency (NSA) created the majority of the algorithms in this group, including SHA-0 (released in 1993), SHA-1 (released in 1995), SHA-2 (released in 2001) and SHA-3 (released in 2014). SHA-2 and SHA-3 are mostly suggested for use in blockchains and cryptocurrencies to meet current security requirements" [24]. The SHA256 is a class of the SHA-2 algorithm and creates a

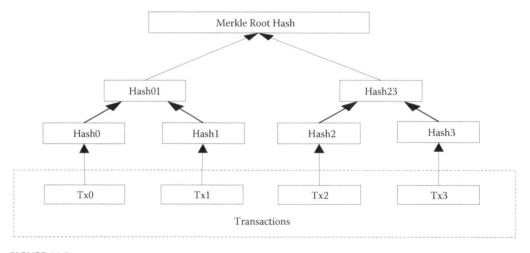

FIGURE 11.5
Merkle tree.

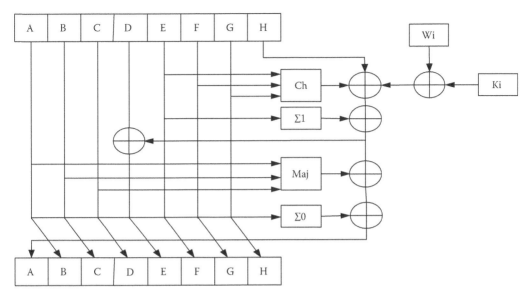

FIGURE 11.6
Core structure of SHA256.

256-bit message digest. The design and implementation of SHA-256 is described in detail in the reference [25]. The structure of each round in SHA-256 is depicted in Figure 11.6. SHA-256 utilizes the majority function represented as equation (11.1), the conditional function represented as the equations (11.2), the rotate functions represented as the equations: (11.3) and (11.4). Each function operates on 32-bit buffers denoted by the letters A, B, C, D, E, F, G, and H. Each function produces a new 32-bit word as a result.

$$Maj\,(A,\,B,\,C) = (A \wedge B) \oplus (A \wedge C) \oplus (B \wedge C) \tag{11.1}$$

$$Ch\,(E,\,F,\,G) = (E \wedge F) \oplus (\neg E \wedge G) \tag{11.2}$$

$$\textstyle\sum_0 (A) = (A \gg 2) \oplus (A \gg 13) \oplus (A \gg 22) \tag{11.3}$$

$$\textstyle\sum_1 (E) = (E \gg 6) \oplus (E \gg 11) \oplus (E \gg 25) \tag{11.4}$$

The usages of hash function used in blockchains can be divided into seven categories such as "proof-of-work, address generation, block generation, Merkle-tree model, message digest in digital signatures, pseudorandom number generation and bridge components" [24]. The first three became famous recently as a result of the advent of cryptocurrencies and blockchain, whereas the final four are fairly common and popular even before the beginning of blockchains.

11.4.3 Digital Signature

The digital signature scheme usually includes two parts: a signing process and a verifying process. The signing process produces a digital signature on the message by using the signer's private key which is kept secret. The verifying process verifies the digital signature of the message by using the signer's public key which is distributed to the public. As a result, the individual who has to verify the signature may do so quickly. The digital signature is usually utilized for ensuring authentication, non-repudiation, identity verification, and integrity verification [22]. A digital signature is used to communicate and trust between nodes in a distributed network of blockchains.

In a cryptocurrency system, digital signatures are used in transactions to prove that the spender possesses the cryptotoken. The cryptotoken owners provide digital signatures in their new transactions using their corresponding private key to verify that they possess the cryptotokens. This process creates a chain of digital signatures. This process hashes the content of the prior transaction and the address of the next owner. The data which is digitally signed with its own private key is added to the end of the transaction list and transmitted to the receiver. The receiver must check the received information in order to prove the prior owner's information, and then verify the owner of the transaction using their public key. The present owner, prior owner, and next owner of the cryptotoken are all recorded in each transaction on the blockchain. This creates a chain of ownerships for each transaction. As a result, the entire transaction of cryptotoken may be traced back, thus successfully preventing double payment and faked transactions [23].

Figure 11.7 depicts the demonstration of signing and verifying processes in each transaction of cryptotoken. For example, suppose that Owner(2) and Owner(3) make a payment transaction. When Owner(2) wants to pay Owner(3) 50 BTC, BTC amount and its source are recorded on the transaction list first. The Owner(2)'s 50 BTC are from the Owner(1), and thus, to achieve the successful payment transaction of the Owner(2) to the Owner(3), it is required to record the BTC source, the payment amount, and the Owner(2)'s digital signature.

The digital signature scheme currently used in the most of blockchains is "Elliptic Curve Digital Signature Algorithm (ECDSA)" [24]. The specific curve used by blockchains for ECDSA is secp256k1, and it is an elliptic curve that has been specified by NIST [26]. ECDSA is mostly implemented for blockchains, as the fact that its key sizes are very small and it has better computational efficiency than RSA. The design and implementation of ECDSA is described in detail in the reference [26]. ECDSA has three segments: key generation, signing, and verifying as shown below.

Key Generation. Owner(1) and Owner(2) agree to select the point P with order n as a base point. Owner(1) calculates their private key and public key as in the following steps.

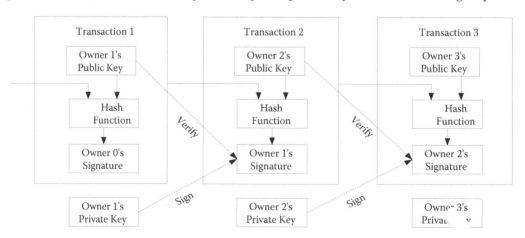

FIGURE 11.7
Signing and verifying for transactions.

- Select an integer $d \in [1, n-1]$ and it is defined as a private key.
- Calculate $Q = d \times P$ and the point Q is defined as a public key.

Signing. Owner(1) calculates the signature as the following steps to sign the message with her private key d.

- Select a random integer$k \in [1, n-1]$.
- Calculate:$R = k \times P = (x_1, y_1)$.
- Calculate: $r = x_1 \bmod n$.
- Calculate: $h = h(m)$ as the hash value of the signing message.
- Calculate: $s = k^{-1}(h + rd) \bmod n$.

After computing steps for signing, Owner(1) sends (r, s) to Owner(2) as their signature of the signing message.

Verifying. Owner(2) receives (r,s) as the signature of the signing message and calculates the following steps to verify the signature with Owner(1)'s public key Q.

- Calculate: $w = s^{-1} \bmod n$.
- Calculate: $h = h(m)$ as the hash value of the same signing message.
- Calculate: $u_1 = h \times w \bmod n$.
- Calculate: $u_2 = r \times w \bmod n$.
- Calculate: $u_1 P + u_2 Q = (x_0, y_0)$.
- Calculate: $v = x_0 \bmod n$

After computing steps for verifying, Owner(2) accepts the signature since $v = r$.

Blockchains make use of a variety of digital signature schemes that have the ability that supports anonymity for the signer. "Aggregate signature, group signature, ring signature, blind signature, and proxy signature are the anonymous signature schemes used in blockchains" [27].

- *Aggregate Signature* – A digital signature that utilizes the aggregation function is known as an aggregate signature. Given n digital signatures on n distinct messages from n various users, aggregation means that all of these signatures may be summarized into a single signature. The aggregate signatures significantly decrease the cost of signature verification and signature storage. The computing and communication costs may be minimized in practice. It is most commonly utilized in situations with limited bandwidth and storage capacity [28].
- *Group Signature* – A group signing method enables members of a group to stand for group signed messages anonymously. "It enables the security aspects such as integrity and reliability, unforgeability, anonymity, traceability, unlinkability, no framing, unforgeable tracing verification, and coalition resistance" [29]. The execution performance of group signature scheme becomes essential due to the various situations in blockchain applications. The group signature length, the public key size, the signature creation time, and the verification time may all be used to assess the execution performance of group signature scheme [30,31].

- *Ring Signature* – A ring signature system combines the public keys of all users on a set with a single signer's private key. It might be utilized in anonymous payment systems or transactions that require anonymity. It enables the security aspects such as unforgeability and anonymity. It does not need the involvement of a trustworthy third party or a group administrator [32].

- *Blind Signature* – A blind signature scheme is frequently used in the privacy-related protocols in which the message author and the signer are not the same person [33]. Blind signature schemes support complete unlinkability and are utilized in digital currency systems that are anonymous. Blind signatures must meet the additional two criteria in addition to the common digital signature conditions. (1) The signer signs the message that is not visible to him. (2) The signed message is not traceable [33].

- *Proxy Signature* – "It is a signature scheme that the original signer delegates his signing authority to the proxy signer, and then the proxy signer creates a valid signature on the behalf of original signer" [34]. "The proxy signature has a direct form as compared to the continuous execution of typical digital signature schemes. In the verification process, the verifier does not require the public key of any user other than the original signer" [27]. Therefore, its computing cost is lower than running a typical signature scheme.

11.4.4 Encryption

In blockchains, data encryption is a key technique. The concept of a confidential transaction was initially introduced to safeguard the privacy of clients. Asymmetric encryption is a type of encryption that is included into the blockchain to provide security and ownership verification. "Asymmetric encryption usually includes two ciphers in the process of encryption and decryption, which use public key and private key respectively" [35]. The mathematical stochastic transformation process of converting the readable format plaintext to the unreadable format ciphertext using a public key is referred to as encryption. The reverse transformation process of converting the unreadable format ciphertext to the readable format plaintext using a private key is referred to as decryption. The public key can be shared with others, but the private key remains secret. The most common encryption method utilized in blockchain applications is the asymmetric encryption based on the curve secp256k1 defined in "Standards for Efficient Cryptography (SEC)" [23]. Technical details and properties of secp256k1 are described in detail in the reference [36].

Elliptic Curve ElGamal Encryption Scheme – Mathematical computing for key generation, encryption, and decryption using elliptic curve is described below [37].

Key Generation. User(1) and User(2) agree to select the point P with order n as a base point. User(2) calculates their private key and public key as in the following steps.

- Select an integer $d \in [1, n - 1]$ and it is defined as a private key.
- Calculate $Q = d \times P$ and the point Q is defined as a public key.

Encryption. User(1) selects the point M as a message to encrypt with the public key Q of the User(2). The cipher text is calculated as following steps.

- Select a random integer $r \in [1, n - 1]$.
- Calculate: $C_1 = r \times P$.
- Calculate: $C_2 = M + (r \times Q)$.

After computing steps for encryption, User(1) sends the points C_1 and C_2 to User(2) as cipher texts.

Decryption. User(2) receives the points C_1 and C_2 as cipher texts to decrypt with their private key d. The message is calculated below.

- Calculate: the message $M = C_2 - (d \times C_1)$.

After computing steps for decryption, User(2) receives the message sent by User(1).

The data on the blockchain is totally open and transparent. Thus, the transaction privacy protection of the blockchain poses security issues. Modern asymmetric encryption technologies and the safe nature of blockchain transactions keep cryptocurrencies secure. Private keys are used by cryptocurrency owners to prove that they are the rightful owners of their tokens.

The following encryption types are also used in some blockchain systems for better privacy and security.

- *Homomorphic Encryption* – Homomorphic encryption is a kind of cryptographic technique that utilizes an additive isomorphism and a multiplicative isomorphism. It computes on encrypted data. It utilizes proxy reencryption technique to prevent the selected ciphertext from being attacked, and it obtains the same result from the calculation of the original data. This method has following steps: (1) create a couple of keys for homomorphic encryption, (2) keep the private key in the hands of a trustworthy third party, (3) transmit it to the blockchain user who is responsible for the transaction, and (4) encrypt using the corresponding public key. As a result, the data security of cryptocurrency transactions can achieve greater privacy protection. "The benefit of this method is that the transaction data can be encrypted using an encryption algorithm before being submitted to the blockchain network; the data exists in a secret text format, and even if an attacker gains access to it, he will not expose any of the user's private information" [38]. For better control and privacy, smart contracts implemented in Ethereum utilize homomorphic encryption on the transaction data stored in the blockchain [39]. Homomorphic encryption schemes are described in detail in the reference [40].

- *Attribute-Based Encryption* – Attribute-based encryption is a kind of asymmetric encryption in which the ciphertext and the user's private key are both determined by attributes. In this system, if the set of the attributes of user key match the attributes of the ciphertext, the encrypted data may be decrypted using the user's private key. To implement attribute-based encryption, there is no requirement for a central authority. In a decentralized network like a blockchain, several authorities can work together to achieve the same goal. The collusion-resistance is a significant security feature of attribute-based encryption. It protects users with different global IDs from colluding together [14,41,42]. For an attribute-based encryption, software implementation is described in the reference [43].

11.4.5 Commitment Protocol

A commitment protocol is a cryptographic technique in which two users play a two-phase game with a digital sealed envelope. "In the first phase, known as the commit phase, user A sends user B a commitment for committing an important secret in the sent envelop. In the second phase, known as the open phase, user A opens the commitment to prove that user A did not cheat user B in the commit phase. The commitment meets two security requirements: hiding and binding. Hiding guarantees that user B cannot tear the envelope for seeing the secret before the open phase, while binding ensures that user A cannot modify what was sealed in the sent envelop after the commit phase" [27,44].

In blockchains, the commitment protocol is also highly beneficial to hide amounts of any transaction. Pedersen commitment is a well-known protocol for commitment, and its variations are mostly utilized in blockchains for the purpose of the confidential trans-action [45,46]. The application of Pedersen commitment for confidential transactions is described in the reference [47].

Pedersen Commitment Using Elliptic Curve – Suppose that P and Q are the points which are randomly generated from a generator point G with order n on an elliptic curve, r is a random integer called the blinding factor, and m is an integer that encodes the hidden message. The elliptic curve scalar multiplication operation is used to form m.P and r.Q. R is a point on the elliptic curve formed by combining m.P and r.Q with the elliptic curve addition operation. The point R is a Pedersen commitment. By revealing both the mes-sage m and the blinding factor r, the Pedersen commitment is opened. The verification of a Pedersen commitment given m, r, and R simply checks that pedersen_commit(m, r) returns the same point R. The pedersen_commit(m, r) is implemented based on additive homomorphism. To form the point R3 = R1 + R2, two Pedersen commitments R1 = m1.P + r1.Q and R2 = m2.P + r2.Q are combined using elliptic curve point addition. The Pedersen commitment to the integer message m3 = m1 + m2 and the new blinding factor r3 = r1 + r2 is represented by the point R3 such that R3 = m3.P + r3.Q [48].

11.4.6 Zero-Knowledge Proof

Zero-knowledge proof is a kind of cryptographic technique with strong privacy-preserving features. The basic concept is that "a certifier can verify to a verifier that an assertion is correct without supplying the verifier with any relevant information. If the certifier and the verifier have a common reference string, it is possible to achieve com-putational zero-knowledge without the requirement for communication between the certifier and the verifier" [22,24,49,50]. The general ideal of a zero-knowledge proof is described in the reference [51].

All account balances are encrypted and saved in a block in a blockchain application. When a user transmits cryptocurrency to another user, he may easily utilize zero-knowledge proof to show that he has sufficient funds for the cryptocurrency transfer while concealing his account balance [14]. The application of zero-knowledge proof for confidential transactions is described in the reference [52].

Zero-Knowledge Proof Using Elliptic Curve – PROVER and VERIFIER agree a generator point G on an elliptic curve with order n. They both know the point P on the curve. PROVER declares he knows $\beta \in n$ such that $P = \beta . G$. They want to prove this fact to VERIFIER without opening β. Protocol steps are as followings [53].

- PROVER selects a random integer $\lambda \in n$ and calculates the point $Q = \lambda . G$.
- PROVER sends the point Q to VERIFIER.
- VERIFIER calculates $\theta = HASH(G, P, Q, \alpha)$ where α is a random seed and sends θ to PROVER.
- PROVER calculates $\phi = \lambda + \theta . \beta \pmod{n}$ and sends ϕ to VERIFIER.

- VERIFIER checks that
- $R = \phi.\, G - \theta.\, P = (\lambda + \theta.\, \beta).\, G - \theta.\, \beta.\, G = \lambda.\, G + \theta.\, \beta.\, G - \theta.\, \beta.\, G = \lambda.\, G = Q.$

11.4.7 Nonce

Nonce is known as number once in cryptography. It is an arbitrary number that can be utilized just once in a cryptographic transaction. It may be a random or pseudo-random number generated as part of an authentication process to prevent replay attacks on previous communications. It can also be used in cryptographic hash algorithms and as an initialization vector [54,55].

In cryptocurrency mining, "a nonce is an arbitrary number that miners adjust to modify the header hash and create a hash that is less than or equal to the target hash value set by the network" [54]. It is appended to a hashed block in a blockchain when a block is rehashed to achieve the difficulty level requirements. Miners attempt to calculate the nonce. When a solution is identified, the miner is rewarded with cryptocurrency [56,57].

11.5 Consensus Techniques

11.5.1 Introduction

Consensus techniques, which do not need third-party verification, allow participants to validate transactions and prevent malicious actors from inserting fake and fraudulent information. A consensus protocol is in charge of ensuring that on a blockchain network, all nodes are in sync with one another. Consensus refers to the fact that the same state of a blockchain is agreed upon by all nodes in a network, allowing it to function as a self-auditing system. It serves as the foundation for blockchain technology, and has two important features to operate. First, a blockchain is allowed to be updated while guaranteeing that every block in the blockchain is correct and participants stay motivated. Second, it prohibits a single entity from derailing or controlling the whole blockchain system. Consensus aims to ensure that just one chain is utilized and followed. When verifying a block and the transactions within it, consensus has a set of rules that must be followed by the blocks in the nodes. Consensus rewards and incentivizes network participants who maintain a blockchain. Consensus is usually planned to be hard to imitate or duplicate by being enormously costly to work out. Depending on the blockchain, the consensus techniques may differ.

11.5.2 Proof of Work (PoW)

PoW is the first consensus technique for validating cryptocurrency transactions. It is designed for permissionless public ledgers that relies on the computational capabilities of the systems in the node. PoW is a type of cryptographic zero-knowledge proof in which the prover proves the verifiers that a particular amount of computing effort has been expended. Verifiers can easily confirm this expenditure with little effort. In other words, PoW is a cryptographic puzzle that is difficult or expensive to make but can be easily validated by the others. Mining is the name given to the PoW process, and miners are the nodes that participate in it. Miners work on difficult cryptographic puzzles that need a lot of computing power. Finally, the difficulty of these puzzles is affected by how quickly blocks are mined. The first miner who solves the puzzle gets to make a block and is rewarded for doing so. A PoW is described in detail in the references [14,58]. This approach is used to prevent users from "double-spending" their coins, denial-of-service attacks, and spam [59,60].

PoW is a computer algorithm that cryptocurrencies like Bitcoin, Ethereum, Litecoin, and others utilize to establish a consensus for adding a specific block to the blockchain. As shown in Figure 11.8, miners use the SHA-256 as a PoW function to solve difficult math problems and add blocks to the blockchain. The hash function produces a specific type of data that can be used to verify that a great deal of work has been completed.

11.5.3 Proof of Stake (PoS)

PoS is the second consensus technique for validating cryptocurrency transactions. It is most often used after PoW. It was first introduced by Peercoin. In this technique, cryptocurrency owners can stake their tokens, offering the chance to check new transaction blocks and add them to the blockchain. To determine who has a chance to build a new block in the blockchain, this method utilizes a randomized mechanism and there is no cryptographic puzzle to be solved. The goal is to minimize centralization of mining centers and provide all miners a chance to validate. To become validators, minners may utilize their tokens by staking them, meaning they put their money in a safe for a particular amount of time in order to generate a new block. The minner who holds the most stakes has the best opportunity of becoming a validator, creating a new block, as shown in Figure 11.9. This method is dependent on how long the cryptotokens have been staked by one minner. The energy of other validators in the network may be reduced by utilizing this consensus technique in case only selected validators can generate a block. In contrast to PoW, the validators that check and validate the block earn their transaction fees. If a

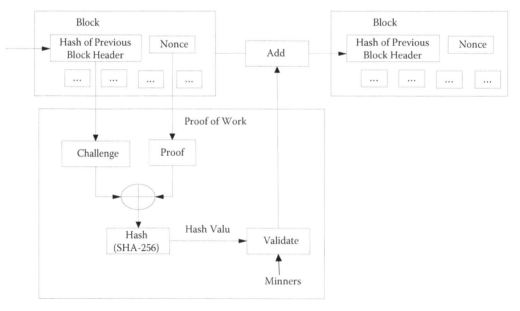

FIGURE 11.8
Proof of work.

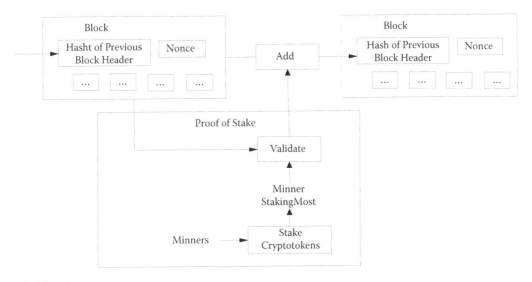

FIGURE 11.9
Proof of stake.

validator makes a mistake during the block generation process, they lose their stakes. PoS is described in detail in the references [14,60]. A 51% attack is a type of risk that occurs on a PoW blockchain. PoS architecture can reduce the likelihood of 51% attacks [59].

11.5.4 Delegated Proof of Stake (DPoS)

DPoS is a fast consensus technique that was first utilized to implement EOS. A delegate refers to a person or organization that wants to generate blocks on the network. The delegates who receive the most votes are given the opportunity to create blocks as shown in Figure 11.10 and are rewarded for doing so. Delegates are rewarded in one of two ways: transaction fees or a particular amount of cryptocurrency produced during inflation. Nodes on network can vote for delegates by staking their cryptocurrencies. The weight of a vote is determined by the stakes. In comparison to other consensuses like PoW and PoS, the DPoS is significantly more efficient in processing transactions [59]. A DPoS is described in detail in the references [14,60]. This approach prevents a transaction from being spent twice in the system.

11.5.5 Proof of Capacity (PoC)

PoC is a consensus technique that utilizes a plotting process. In PoW, miners utilize computing power to decide the right solution but in PoC, solutions are pre-stored in the hard-disk memory. The miners make a plot using the storage data. So, this process is known as plotting. Miners can participate in the block generation process after the storage data has been plotted. If a miner has more capacity, it can store more solutions. As a result, miners with more storage capacity have a higher chance of creating a new block, as shown in Figure 11.11. Burstcoin, Storj, Chia, and SpaceMint are cryptocurrencies that utilize this method [59,60].

11.5.6 Proof of Elapsed Time (PoET)

PoET is a consensus method that uses a fair lottery approach to prevent high resource utilization and energy consumption, making the process more efficient on a blockchain

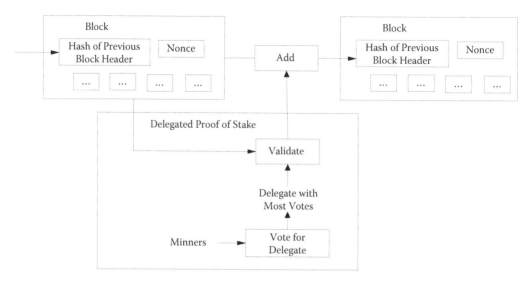

FIGURE 11.10
Delegated proof of stake.

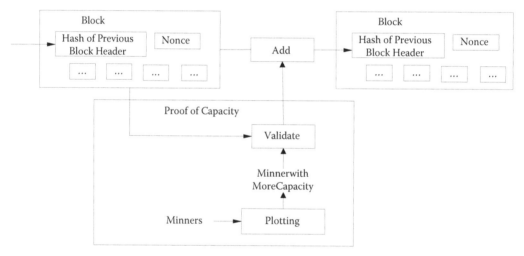

FIGURE 11.11
Proof of capacity.

network. This method selects miners at random and fairly. On a blockchain network, this method decides mining rights and block winners using a randomly generated elapsed time. By selecting a miner, it also determines who gets to create a new block. All nodes in the network are given a random and fair wait time by this process. This technique is dependent on how long the miners have been waiting for the block to be created. The node with the shortest wait time gets to create a new block first, as shown in Figure 11.12. This technique is useful for ensuring that a system does not have duplicate nodes and that the allocated wait time is a random amount [59,60]. PoET is described in detail in the references [61–63].

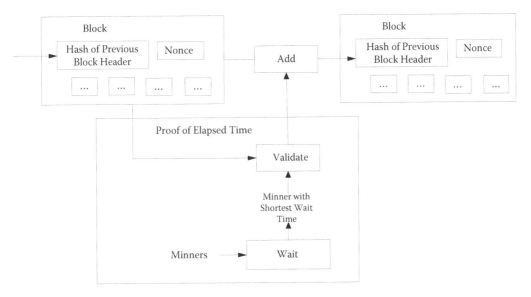

FIGURE 11.12
Proof of elapsed time.

11.5.7 Proof of Activity

Proof of activity is a hybrid method that incorporates both PoW and PoS. Cryptographic puzzles were solved by miner in the same way they did in PoW before switching to PoS. Blocks, rather than transactions, contain templates that include header information as well as the address of the reward for mining. In this technique, the blocks are established by limiting the time it takes to create a block to the shortest amount of time possible for allowing the maximum number of blocks to be added to the chain. As a result, the network is protected from spam transactions, avoiding floods. This approach makes it impossible for the Bitcoin network to be hacked. It also prevents the likelihood of a 51% attack, like in PoS [60].

11.5.8 Proof of Space

Proof of space is a network consensus technique that works in a similar way to PoW. This method validates transactions using disk storage instead of computing resources and rewards miners who have the largest disk space allocated to a block. For anti-spam and denial-of-service attack protection, proofs of space might be employed instead of PoW. Proof of space has also been used to identify malware [64]. Proof of space is described in detail in the reference [65].

11.5.9 Proof of Importance (PoI)

PoI is a consensus technique utilized to decide which network nodes are qualified to add a block to the blockchain and it was first developed by the New Economy Movement (NEM). PoI, which is based on PoS, rewards users who actively transact on the network. PoI requires nodes to invest a certain amount of currency in order to be eligible to generate blocks, and nodes are chosen to create blocks based on a score that quantifies their contribution to the network. Nodes with a higher significance score have a better chance of getting picked to create a block [60,66].

11.5.10 Proof of Ownership (PoO)

PoO is a consensus technique designed for tracking the owners of certain information at a given point in time. Using PoO, both the prover and the verifier can check the ownership of an asset at zero cost. It keeps deceitful records on secondary markets and allows safer P2P transactions. This technique is used to confirm the integrity, date of publishing, and ownership of their products or contracts in businesses. In this approach, only those who have access to private key of the signature may verify they are the owner. A piece of data is always accompanied with a PoO. Certifying the ownership of a piece of data on the blockchain confirms that the data existed at a specific point in time. Cryptographic functions are always used to attach a PoO to a piece of data. This makes it impossible to change the data after it has been certified. If even a single bit of the content is changed, the entire certificate becomes invalid. Because proofs are stored on the blockchain, it is simple to transfer ownership and to track the original issuer of a certificate. Certificates cannot be double-spent. It applies not only to the CodeChain platform, but also to Bitcoin, Ethereum, Cardano, Litecoin, and other cryptocurrencies [60,67,68].

11.5.11 Proof of Burn (PoB)

PoB is an alternative consensus technique for PoS and PoW. PoB uses less energy than PoW. This technique ensures that miners achieve a consensus by burning crypto-currencies. The process of permanently removing cryptocurrencies from circulation is known as burning. PoB-powered blockchains utilize it to validate transactions, in case of the fact that it reduces inflation. PoB is easily verifiable but difficult to undo in comparison to PoW and PoS [69]. PoB validates transactions using virtual mining equipment rather than real ones. As a result, the more coins you have, the more power you have, and vice versa. Higher mining capability increases the speed of finding new blocks. As a result, the miner earns more rewards. PoB is described in detail in the reference [70].

11.5.12 Byzantine Fault Tolerance (BTF)

A Byzantine fault occurs when components in a distributed computing system fail and there is no sufficient information to decide whether a component has failed. The goal of BTF is to be able to resist against system component failures with or without symptoms that prevent other system components from reaching a consensus among them, which is required for correct operation of the system [71]. The NEO platform utilized BTF as a consensus technique. BFT is a system that provides correct service without interruptions. Service interruptions are mostly caused by Byzantine faults such as software flaws, operator errors, and malicious assaults. Therefore, practical Byzantine fault tolerance (PBFT) was developed to overcome a number of issues with existing BFT solutions. It is designed to have a minimal overhead time. PBFT is described in detail in the reference [72].

11.5.13 Proof of Authority (PoA)

PoA is a consensus technique based on identity as a stake to support comparably quick transactions. It was utilized by VeChain. PoA has a core principle that "only validators have the authority to approve new blocks and transactions" [73]. A participating node gains reputation for his identity, and the node may only become a validator after his reputation has reached a high level. PoA is a modified form of PoS and it is seen to be more reliable than PoS. Validators are encouraged to support the transaction process by attaching a reputation to their identities, since they do not want their identities to be associated with a negative reputation. "A validator is not able to approve any two consecutive blocks" [74]. As a result, trust is not centralized. This method removes the risks that come with having anonymous block producers, as can be seen with PoW or PoS.

11.6 Conclusion

Cryptographic techniques described in section (4) and consensus techniques described in section (5) are the recent privacy and security techniques supporting the privacy and security management of cryptocurrency transactions. Based on the privacy and security requirements described in section (3.2), the appropriate privacy and security techniques described in sections (4) and (5) should be chosen and utilized for a complex blockchain system to achieve security properties described in section (3.3). We hope that these techniques support the privacy and security management of cryptocurrency transactions on the blockchain network.

References

[1] Wikipedia, "Bitcoin", https://en.wikipedia.org/wiki/Bitcoin

[2] CoinMarketCap, https://coinmarketcap.com

[3] Conway, L. W., "The 10 Most Important Cryptocurrencies Other Than Bitcoin", https://www.investopedia.com/tech/most-important-cryptocurrencies-other-than-bitcoin (2021)

[4] Network Encyclopedia, "Peer to Peer Network", https://networkencyclopedia.com/peer-to-peer-network-p2p

[5] MYCRYPTOPEDIA, "Timestamp Server", https://www.mycryptopedia.com/bitcoin-explained

[6] Atlam, H. F., A. Alenezi, et al., "Blockchain with Internet of Things: Benefits, Challenges, and Future Directions." *International Journal of Intelligent Systems and Applications* 10 (2018) (6): 40–48

[7] Liang, Y.-C., "Blockchain for dynamic spectrum management", *In book: Dynamic Spectrum Management, Signals and Communication Technology*. Singapore: Springer. (2019)146–212.

[8] Latifaer er-rajy, M. el Ghazouani, et al., "Blockchain: Bitcoin Wallet Cryptography Security, Challenges and Countermeasures." *Journal of Internet Banking and Commerce* 22 (2017): 1–29.

[9] Zhu, L.-H., B.-K. Zheng, M. Shen, et al., "Data Security and Privacy in Bitcoin System: A Survey." *Journal oF Computer Science and Technology* 35 (2020) (4): 843–862.

[10] Guru, D., S. Perumal and V. Varadarajan, "Approaches towards Blockchain Innovation: A Survey and Future Directions." *Electronics* 10 (2021) (10): 1219.

[11] Euromoney Learning, "How Does a Transaction Get into the Blockchain?", https://www.euromoney.com/learning/blockchain-explained/how-transactions-get-into-the-blockchain

[12] Investopedia, "Blockhain Explained", https://www.investopedia.com/terms/b/blockchain.asp

[13] Mistry, N., "An Introduction to Bitcoin, Elliptic Curves and the Mathematics of ECDSA", https://www.slideshare.net/NikeshMistry1/introduction-to-bitcoin-and-ecdsa

[14] Zhang, R., R. Xue and L. Liu, "Security and Privacy on Blockchain." *ACM Computing Surveys* 52 (2019) (3): 1–34.

[15] Investopedia, "Double-Spending", https://www.investopedia.com/terms/d/doublespending.asp

[16] Ali, A. and M. M. Afzal, "Confidentiality in blockchain." *International Journal of Engineering and Science Invention* 7 (2018) (1): 50–52.

[17] Wang, Y. and A. Kogan, "Designing confidentiality-preserving blockchain-based transaction processing systems." *International Journal of Accounting Information Systems* 30 (2018):1–18.

[18] Hulya Boydas Hazar, "Anonymity in cryptocurrencies". In: Bilgin M. H., Danis H., Demir E. eds.: *Eurasian Economic Perspectives. Eurasian Studies in Business and Economics*, vol 14/1, Cham, Springer. (2020) 171–178.

[19] Blockchain.com, "Network difficulty", https://www.blockchain.com/charts/difficulty

[20] Investopedia, "Cryptocurrency Difficulty", https://www.investopedia.com/terms/d/difficulty-cryptocurrencies.asp

[21] CLOUDFLARE, "Cryptocurrency DDoS Attacks", https://www.cloudflare.com/learning/ddos/cryptocurrency-ddos-attacks/

[22] Forouzan, B. A., *Cryptography and Network Security*. Boston: McGraw-Hill, 2008.

[23] Zhai, S., Y. Yang, et al., "Research on the Application of Cryptography on the Blockchain." *Journal of Physics: Conf. Series* 1168 (2019) (032077).

[24] Wang, L., X. Shen, et al., "Cryptographic primitives in blockchains." *Journal of Network and Computer Applications* 127 (2019): 43–58.

[25] NIST, "Secure Hash Standard", *FIPS PUB 180-2*, https://csrc.nist.gov/publications/detail/fips/180/2/archive/2002-08-01

[26] Johnson, D. and A. Menez, "The Elliptic Curve Digital Signature Algorithm ECDSA", https://citeseerx.ist.psu.edu/viewdoc/download?doi=10.1.1.472.9475&rep=rep1&type=pdf

[27] Fang, W., W. Chen, et al. "Digital signature scheme for information non-repudiation in blockchain: A state of the art review." *Journal on Wireless Communications and Networking* 56 (2020): 1–15.

[28] Boneh, D., C. G. Ben, et al., "Aggregate and Verifiably Encrypted Signatures from Bilinear Maps," in Proceedings of International Conference on the Theory and Applications of Cryptographic Techniques. Springer, Berlin, Heidelberg, 2003, pp. 416–432.

[29] Manulis, M., et al., "Group Signatures: Authentication with Privacy", *Bundesamt für Sicherheit in der Informationstechnik* (2012), https://www.bsi.bund.de/SharedDocs/Downloads/EN/BSI/Publications/Studies/GruPA/GruPA.pdf?__blob=publicationFile

[30] Chaum, D. and E.V. Heyst, "Group Signatures," In Proceedings of Advances in Cryptolog. Springer, Berlin, Heidelberg, 1991, pp. 257–265.

[31] Wikipedia, "Ring Signature", https://en.wikipedia.org/wiki/Ring_signature

[32] Rivest, R. L., A. Shamir and Y. Tauman, "How to Leak a Secret," In Proceedings of Advances in Cryptology —ASIACRYPT. Gold Coast, Australia, 2001, pp. 552–565.

[33] Wikipedia, "Blind Signature", https://en.wikipedia.org/wiki/Blind_signature

[34] Mambo, M., K. Usuda and E. Okamoto, "Proxy Signatures for Delegating Signing Operation," In Proceedings of the 3rd ACM Conference on Computer and Communications Security. New Delhi, India, 1996, pp. 48–57.

[35] Springer Link, "Asymmetric Encryption", https://link.springer.com/referenceworkentry/10.1007%2F978-0-387-39940-9_1485

[36] Bitcoin Wiki, "Secp256k1", https://en.bitcoin.it/wiki/Secp256k1

[37] Aung, T. M. and N. N. Hla, "A new technique to improve the security of elliptic curve encryption and signature schemes," In *Future Data and Security Engineering*. Springer LNCS, vol. (11814), 2019.

[38] Chen, J. and F. You, "Application of Homomorphic Encryption in Blockchain Data Security," EITCE 2020: Proceedings of the 2020 4th International Conference on Electronic Information Technology and Computer Engineering, 2020, pp. 205–209.

[39] Yan, X., Q. Wu and Y. Sun, "A Homomorphic Encryption and Privacy Protection Method Based on Blockchain and Edge Computing." *Wireless Communications and Mobile Computing* (2020): 1246–1257.

[40] Acar, A., H. Aksu, et al., "A Survey on Homomorphic Encryption Schemes." *ACM Computing Surveys* 51 (2018) (4): 1–35.

[41] Wikipedia, "Attribute-based Encryption", https://en.wikipedia.org/wiki/Attribute-based_encryption

[42] Lewko, A. and B. Waters, [n. d.]. "Decentralizing Attribute-based Encryption." *EUROCRYPT* (2011): 568–588. 2006-CS-001-000001-02

[43] Zavattoni, E., et al., "Software Implementation of an Attribute-Based Encryption Scheme." *IEEE Transactions on Computers* 64 (2015) (5): 1429–1441.

[44] Damg°ard, I. and J. B. Nielsen, "Commitment Schemes and Zero-Knowledge Protocols", Aarhus University, BRICS, (2006). https://courses.cs.ut.ee/2006/crypto-seminar-fall/files/kamm1.pdf

[45] Pedersen, T. P., "Non-interactive and Information-theoretic Secure Veriable Secret Sharing," Advances in Cryptology - CRYPTO '91, 11th Annual International Cryptology Conference, Santa Barbara, California, USA, August 11–15, 1991, Proceedings, Vol. 576 of Lecture Notes in Computer Science, Springer, 1991, pp. 129–140.

[46] Hanser, C. and D. Slamanig, "Revisiting Cryptographic Accumulators, Additional Properties and Relations to Other Primitives," Topics in Cryptology- CT-RSA 2015, The Cryptographer's Track at the RSA Conference 2015, San Francisco, CA, USA, April 20-24, 2015. Proceedings, Vol. 9048 of Lecture Notes in Computer Science, Springer, 2015, pp. 127–144.

[47] Nccgroup, "On the Use of Pedersen Commitments for Confidential Payments", https://research.nccgroup.com/2021/06/15/on-the-use-of-pedersen-commitments-for-confidential-payments

[48] Findora, "Pedersen Commitment with Elliptic Curves", https://findora.org/faq/crypto/pedersen-commitment-with-elliptic-curves/

[49] Wikipedia, "Zero-Knowledge Proof", https://en.wikipedia.org/wiki/Zero-knowledge_proof

[50] Damg°ard, I. and J. B. Nielsen, "Commitment Schemes and Zero-Knowledge Protocols", Aarhus University, BRICS, (2006). https://courses.cs.ut.ee/2006/crypto-seminar-fall/files/kamm1.pdf

[51] ALTOROS, "A Zero-Knowledge Proof: Improving Privacy on a Blockchain", https://www.altoros.com/blog/zero-knowledge-proof-improving-privacy-for-a-blockchain

[52] Nccgroup, "On the Use of Pedersen Commitments for Confidential Payments", https://research.nccgroup.com/2021/06/15/on-the-use-of-pedersen-commitments-for-confidential-payments

[53] Chatzigiannakis, I., A. Pyrgelis, et al., "Elliptic Curve Based Zero Knowledge Proofs and their Applicability on Resource Constrained Devices," 2011 IEEE Eighth International Conference on Mobile Ad-Hoc and Sensor Systems, 2011, pp. 715–720.

[54] Wikipedia, "Cryptographic Nonce", https://en.wikipedia.org/wiki/Cryptographic_nonce

[55] MYCRYPTOPEDIA, "Bitcoin Nonce Explained", https://www.mycryptopedia.com/bitcoin-nonce-explained

[56] Investopedia, "Nonce", https://www.investopedia.com/terms/n/nonce.asp

[57] Wikipedia, "Proof of Work", https://en.wikipedia.org/wiki/Proof_of_work

[58] Nakamoto, S. "Bitcoin: A Peer-to-Peer Electronic Cash System", www.Bitcoin.Org, 9 (2008).

[59] Hackernoon, "Different Blockchain Consensus Mechanisms", https://hackernoon.com/different-blockchain-consensus-mechanisms-d19ea6c3bcd6 (2018).

[60] Aggarwal, S. and N. Kumar, "Cryptographic Consensus Mechanisms", *Advances in Computers* 149(2020): 270–299.

[61] Investopedia, "Proof of Elapsed Time", https://www.investopedia.com/terms/p/proof-elapsed-time-cryptocurrency.asp

[62] Chen, L., et al., "On security analysis of Proof-of-Elapsed-Time (PoET)". In: Spirakis P. and P. Tsigas (eds.) *Stabilization, Safety, and Security of Distributed Systems. SSS 2017. Lecture Notes in Computer Science*, vol 10616. Cham, Springer. (2017) 282–297.

[63] Bowman, M., et al., "On Elapsed Time Consensus Protocols", https://eprint.iacr.org/2021/086.pdf

[64] Wikipedia, "Proof of Space", https://en.wikipedia.org/wiki/Proof_of_space

[65] Dziembowsk, S., et al., "Proof of Space", https://eprint.iacr.org/2013/796.pdf

[66] MYCRYPTOPEDIA, "Proof of Importance Explained", https://www.mycryptopedia.com/proof-of-importance

[67] Sepa, "Proof of Ownership", https://www.sepa.at/?p=1

[68] Bitcoin Wiki, "Proof of Ownership", https://en.bitcoin.it/wiki/Proof_of_Ownership

[69] Alexandria, "Proof of Burn", https://coinmarketcap.com/alexandria/glossary/proof-of-burn-pob

[70] Karantias, K., A. Kiayias and D. Zindros, "Proof-of-Burn", https://eprint.iacr.org/2019/1096.pdf

[71] Wikipedia, "Byzantine Fault", https://en.wikipedia.org/wiki/Byzantine_fault

[72] Castro, M. and B. Liskov, "Practical Byzantine Fault Tolerance", In Proceedings of the Third Symposium on Operating Systems Design and Implementation, 1999, pp. 173–186.

[73] Alexandria, "Proof of Authority", https://coinmarketcap.com/alexandria/glossary/proof-of-authority-poa

[74] Wikipedia, "Proof of Authority", https://en.wikipedia.org/wiki/Proof_of_authority

12

Secured Blockchain-Based Database for Storing Sensitive Information for Access Control IoT Systems

Smarta Sangui, Suman Das, and Swarup Kr Ghosh

Sister Nivedita University, Kolkata, India

CONTENTS

12.1 Introduction

IoT devices have established themselves as an integral part of our daily life and industry [1]. There are billions of IoT devices deployed all over the world. All these devices exchange a large amount of data that exposes them to various security concerns. Since most IoT devices have low computation power, high-end security algorithms cannot be used effectively and efficiently [2]. So these systems are often subjected to tampering and security breaches. Access management security devices are used widely in many places such as banks, laboratories, research facilities, etc., which make our work easier, but with that new threats are also emerging. Sensitive data like biometric details, confidential information, other personal information, etc. need special attention mostly in the case of access control systems [3]. For example, hackers could intercept network traffic between mobile apps and smart locks. To solve this problem we have introduced a blockchain-based database that would be immutable, decentralized and highly secured [4,5]. A lot of studies were done related to the implementation of blockchain in IoT to solve various problems like security and management. Mengmei et al. [6] used Ethereum as their blockchain platform to manage IoT devices.

We have developed a blockchain based on the Ethereum platform [7] and smart contract [8] that is used to store biometrics and other access information. The smart contract is deployed with the help of Solidity (programming language) [9]. Access of a person will be verified over the blockchain using the Ethereum client. If verified, then the access will

DOI: 10.1201/9781003203957-15

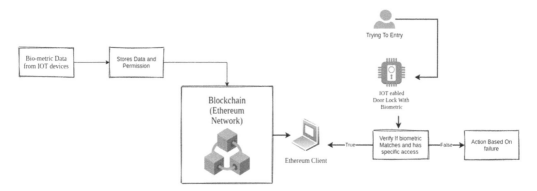

FIGURE 12.1
Block diagram of the proposed model.

be granted. We have mainly focused on the part of verification and storage of sensitive information. Any other loophole is beyond this study. Motivated by Wang et al. (2019) [10], we have implemented blockchain technology that can be useful in various secured places like homes (smart locks), bank vaults, labs, data centers etc. Figure 12.1 shows the block diagram of the proposed blockchain-based security model.

12.2 IoT and Blockchain

Kevin Ashton coined the term *IoT* in 1999 with reference to supply chain management [11]. The Internet of Things (IoT) is a system of interconnected devices that fetc.h data from the local environment and share it to process, store, and analyze. All smart devices, embedded systems, and sensors are IoT [12]. IoT networks are showing a rapid increase in demand and size. Cisco predicted that there will be 500 billion IoT devices by the year 2030. Yet centralized systems are getting expensive due to high deployment and main-tenance overheads [13]. Along with these pre-existing problems trust and security is also a matter of concern in these traditional systems as data can be forged and modified [14]. Proper measures should be introduced to enhance the security and trust of the systems that are often a gateway to highly confidential information.

Blockchain is a kind of tamper-resistant data structure where the data is grouped into a chain of blocks [15]. It is distributed, decentralized, and immutable [16]. Distributed ar-chitecture means that data is not stored in a single node. In this peer-to-peer network, each node in the network keeps a copy of the blockchain so the failure of a single system does not affect the whole network. Decentralization implies no single user has control over the data [17]. It also eliminates the requirement of a third-party validation for transactions, which enhances trust and privacy. Blockchain security methods use public-key cryptography that gives its owners access to their digital assets [18]. Immutability helps to avoid any un-authorised alteration of records. To pass a change in a blockchain, we need to have the approval of 50% of the network. That means if someone has control over 51% of the net-work they can control the whole network. This is also known as the 51% attack [19].

A block contains the cryptographic hash of its data and the hash of the previous block and so on forming a chain. It is similar to the linked list data structure [16]. So if data is

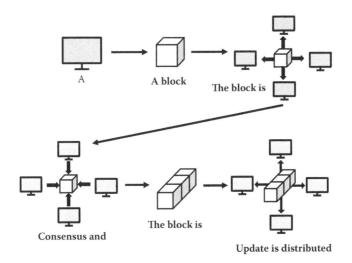

FIGURE 12.2
Blockchain architecture.

modified in one block, the hash value changes in every subsequent block which provides evidence of tampering [20]. Hashing is a mathematical technique that is simple to perform, but very difficult to reverse. Figure 12.2 shows a simple blockchain transaction. Satoshi Nakamoto (alias) introduced blockchain in 2008 as a digital ledger of the famous cryptocurrency Bitcoin [21]. However, since its formation, blockchain has evolved to have multiple use cases in the present day, ranging from banking and payments, supply chain management, audit and regulatory compliance, health care, real estate, voting, and many more [22].

Blockchain is getting popular and many applications are using blockchain to solve real-world problems as it provides security, trust, and privacy. The three most prominent types of blockchains are public, private, and consortium blockchains [23]. In a public blockchain, anyone can read and access the information and developers do not bother the users. All the information is public by default. It is primarily applied to Ethereum [7], Bitcoin [21], Hyperledger [24], and other fields. In the case of a private blockchain, the write permission is only available to an individual or organization, and the read and other permissions are restricted. A private blockchain provides superior security in terms of data privacy and faster transaction speeds in comparison to public blockchains [25]. A consortium blockchain gives us what we may call "partial decentralization," the users can trade and consult, but they won't be able to verify the transaction or publish smart contracts without the permission of the consortium [26]. Blockchains also use varied consensus algorithms depending on the use case. The three most popular consensus are proof of work (PoW), proof of stake (PoS), and practical Byzantine fault tolerance (PBFT).

12.3 Ethereum Blockchain and Smart Contracts

Ethereum is a distributed computing platform, unlike the traditional client-server model [7]. Ethreum mainly comprises blockchain, network trie roots, and trie data structures.

The blockchain logs the network's state at certain times after transactions. These logs are stored in Merkle Patricia trees, which own a hash [27]. In Merkle Patricia trees, every node has a hash value. The hash of each node is determined by the SHA3 hash value of its contents. This hash acts as the key that connects to the node. LevelDB [28] is used as the database layer where the state data is stored. Smart contracts and user account data are stored in the trie data structures. Smart contracts are programmed in Turing complete languages [29] like Solidity, an object-oriented language based on C++, Python, JavaScript, and PowerShell.

Ethereum, NXT, and Hyperledger Fabric were also introduced as blockchain-based cryptocurrencies. Unlike the much popular Bitcoin, they can use smart contracts [9]. The idea of smart contracts was first proposed by Nick Szabo in 1994 [30]. Smart contracts are programs and protocols that are stored in a blockchain that automatically gets executed when relevant events or actions occur according to the contract. They are mostly used to automate the result of the agreement upon fulfilling the conditions so that all stakeholders can be certain of the outcome without any involvement of any intermediary, unlike traditional contracts [31]. It helps prevent tampering as smart contracts are copied to each node of the blockchain. It also helps to reduce human errors and disputes [32].

Though smart contacts developed over the years solve many problems, they still encounter many challenges. In 2016, the Decentralized Autonomous Organization (DAO) smart contract was hacked and manipulated to steal cryptocurrency worth $50 million USD at that time [33] (Figure 12.3).

It is not easy to incorporate blockchain solutions into existing IoT systems due to their low computational resources. And also in a blockchain, each node must have a copy of the blockchain to maintain consistency. Blockchain technology is quite resource-intensive. Scaling them is harder. Several improvements are being made to increase scalability and speed, though much improved yet it is still far from traditional storage methods [34]. On-chain techniques like increasing the block size as in Bitcoin cash [35], compressing the blocks as in compact block relay [36], sharding [37,38], and other improved consensus mechanisms [39,40]. But since access-control IoT systems have to store only a limited amount of data we did not worry about storage. There are many improvements made in this area, like Wang et al. [10] proposed a three-layered hierarchical architecture.

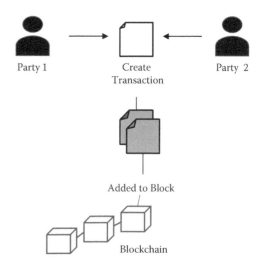

FIGURE 12.3
Smart contract.

12.4 Opportunities

Blockchain has great potential and day by day new solutions are coming up that leverage blockchain to solve problems of trust and security. Scaling a blockchain's security scheme is easier than traditional ones due to its decentralized nature and it also provides strong protection against data tampering and prevents rogue devices [41]. A few salient features of blockchain are listed below that makes it so popular [42].

- Tracking sensor data and preventing unauthorized modification.
- The distributed ledger can be used for device authentication, identification, and safe data transfer.
- Third party is not required for trust.
- Single source failure doesn't affect the system.
- Reduces operational costs as there are no intermediaries.

Though blockchain has its perks and benefits, it is still in its early stages and there are barriers that prevent its wide application in the IoT environment.

One of the challenges is the issue of storage when blockchain is adopted in IoT applications. Usually, IoT devices generate a huge amount of data. Over time, as the blockchain grows, all nodes will require larger storage and better bandwidth to keep the transactions updated [43]. Many studies are done incorporating blockchain with traditional cloud-based IoT architecture.

Since in this paper we are discussing access-control IoT security devices such as biometric locks, we store a limited amount of data. Usually, they store the biometric data of the selected persons who have access. Storing these sensitive data can address the security issues often faced by smart IoT locks. Mohatar et al. [44] discussed using blockchain for biometric template protection.

12.5 Basic Architecture and Implementation

We are using Ethereum blockchain to deploy smart contracts. There are other options available also like Tezos, Quorum, etc. But we will use Ethereum as it is more mature, stable, well documented, has an active community that aids in its maintenance, and provides fine-grained management. We have used Ethereum's smart contract feature extensively for our research. To write a smart contract, we will use the Solidity programming language. Solidity was created by the Ethereum community. As per the Solidity documentation, this language is statically typed and curly braced, which is to be used on an EVM (Ethereum Virtual Machine). The proposed smart contract consists of two sections, one is basic CRUD (create, read, update, and delete) operations for admin and the other part is to be used by the client to verify the user's authenticity via searching the database. The client part is where the IoT edge interacts with a blockchain directly, the access lock part is implemented using an edge device where the user provides their biometric details, either fingerprint, retina scan, or other advanced biometric features. The edge part interacts with Ethereum using the Ethereum client and verifies the user details.

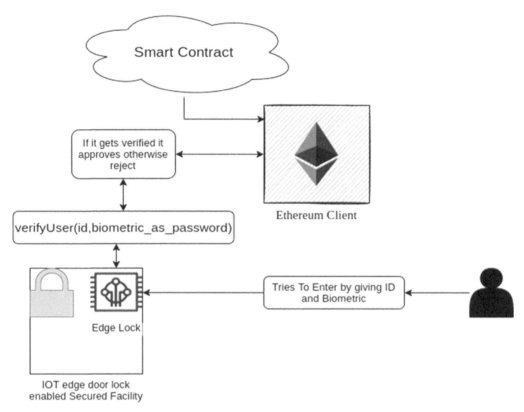

FIGURE 12.4
Smart contact based access management.

If the details get matched, then the edge device allows the user to access the restricted resource. In the admin part, the admin will be responsible for registering users by their identity attribute (ID) and biometric details, updating the required fields e.g ID or biometric details, moderating the permission like giving access to someone or revoking access and deleting someone's account could be performed by admin only using the web or other interfaces created using available open-source Ethereum client. Web3 js is a secure Ethereum client to connect with a local or remote Ethereum blockchain. Figure 12.4 shows the layout of our proposed model.

Now we will look into the core code, which has been written using Solidity version 0.8.4. At first, we need to define the data structure for storage, as shown below.

```
contract SecureStore {
 struct User {
  uint id;
  string name;
  string passwd_biometric;
 }
```

For the admin management, we have to define many CRUD operation functions, to modify the behavior of the smart contract residing inside the Ethereum blockchain.

scenario in the blockchain and IoT world and their advantages and challenges. And we proposed a model to secure the sensitive access control data by using free and open-source resources available. The model is made robust in terms of security with the help of the application of blockchain technology (secured storage) in IoT devices. Although the code is tested and found working, it is not yet tested in a real environment (connect a biometric access control IoT device with proper admin panel) due to lack of resources and the ongoing pandemic.

In the future, we can check the scalability of the architecture, cost, complexity, and efficiency comparisons with other existing technologies. Though real implementation can be challenging at the present, it is a promising research area and further research will solve many drawbacks and reduce complexities and costs.

References

[1] Miller, D., "Blockchain and the Internet of Things in the Industrial Sector." *IT Professional* 20 (2018) (3): 15–18.

[2] Dittmann, G. and J. Jelitto, "A Blockchain Proxy for Lightweight IoT Devices," 2019 Crypto Valley Conference on Blockchain Technology (CVCBT), 2019, pp. 82–85.

[3] Yue, Y., S. Li, P. Legg, F. Li, "Deep Learning-Based Security Behaviour Analysis in IoT Environments: A Survey." *Security and Communication Networks* 2021 (2021), Article ID 8873195: 13 pages.

[4] Fernández-Caramés, T. M. and P. Fraga-Lamas, "A Review on the Use of Blockchain for the Internet of Things." *IEEE Access* 6 (2018): 32979–33001.

[5] Truong, H. T. T., M. Almeida, G. Karame and C. Soriente, "Towards Secure and Decentralized Sharing of IoT Data," 2019 IEEE International Conference on Blockchain (Blockchain), 2019, pp. 176–183.

[6] Huh, S., S. Cho and S. Kim, "Managing IoT Devices using Blockchain Platform," 2017 19th International Conference on Advanced Communication Technology (ICACT), 2017, pp. 464–467, doi: 10.23919/ICACT.2017.7890132.

[7] Buterin, V., Ethereum White Paper, 2013, https://ethereum.org/en/whitepaper/

[8] Buterin, V., A Next-Generation Smart Contract and Decentralized Application Platform, 2014, https://translatewhitepaper.com/wp-content/uploads/2021/04/EthereumOrijinal-ETH-English.pdf

[9] Oliva, G. A., A. E. Hassan and Z. M. Jiang, "An Exploratory Study of Smart Contracts in the Ethereum Blockchain Platform." *Empirical Software Engineering Journal* 25 (2020): 1864–1904.

[10] Wang, G., Z. Shi, M. Nixon and S. Han, "ChainSplitter: Towards Blockchain-Based Industrial IoT Architecture for Supporting Hierarchical Storage," 2019 IEEE International Conference on Blockchain (Blockchain), 2019, pp. 166–175.

[11] Ashton, K., "Internet of Things." *RFID Journal* 22 (2009) (7): 97–114.

[12] Shafique, K., B. A. Khawaja, F. Sabir, S. Qazi and M. Mustaqim, "Internet of Things (IoT) for Next-Generation Smart Systems: A Review of Current Challenges, Future Trends and Prospects for Emerging 5G-IoT Scenarios." *IEEE Access* 8 (2020): 23022–23040.

[13] Khan, M. A. and K. Salah, "IoT Security: Review, Blockchain Solutions, and Open Challenges." *Future Generation Computer Systems* 82 (2018): 395–411.

[14] MacDermott, A., T. Baker and Q. Shi, "Iot Forensics: Challenges for the Ioa Era," 2018 9th IFIP International Conference on New Technologies, Mobility and Security (NTMS), 2018, pp. 1–5.

[15] Nofer, M., P. Gomber, O. Hinz et al. "Blockchain." *Business and Information Systems Engineering* 59 (2017) 183–187.

[16] Syed, T. A., A. Alzahrani, S. Jan, M. S. Siddiqui, A. Nadeem and T. Alghamdi, "A Comparative Analysis of Blockchain Architecture and its Applications: Problems and Recommendations." *IEEE Access* 7 (2019): 176838–176869.

[17] Bodkhe, U. et al., "Blockchain for Industry 4.0: A Comprehensive Review." *IEEE Access* 8 (2020): 79764–79800.

[18] Henry, R., A. Herzberg and A. Kate, "Blockchain Access Privacy: Challenges and Directions." *IEEE Security & Privacy* 16 (2018) (4): 38–45.

[19] Ye, C., G. Li, H. Cai, Y. Gu and A. Fukuda, "Analysis of Security in Blockchain: Case Study in 51%-Attack Detecting," 2018 5th International Conference on Dependable Systems and Their Applications (DSA), 2018, pp. 15–24.

[20] Monrat, A. A., O. Schelén and K. Andersson, "A Survey of Blockchain From the Perspectives of Applications, Challenges, and Opportunities." *IEEE Access* 7 (2019): 117134–117151.

[21] Nakamoto, S., *Bitcoin: Peer-to-Peer Electronic Cash System*, 2008. https://bitcoin.org/bitcoin.pdf

[22] Zīle, K. and R. Strazdiņa, "Blockchain Use Cases and Their Feasibility." *Applied Computer Systems* 23 (2018) (1): 12–20.

[23] Wazid, M., A. K. Das, S. Shetty and M. Jo, "A Tutorial and Future Research for Building a Blockchain-Based Secure Communication Scheme for Internet of Intelligent Things." *IEEE Access* 8(2020): 88700–88716.

[24] Androulaki, E., A. Barger, V. Bortnikov, C. Cachin, K. Christidis, A. De Caro, D. Enyeart, C. Ferris, G. Laventman, Y. Manevich et al., "Hyperledger Fabric: A Distributed Operating System for Permissioned Blockchains," Proceedings of the Thirteenth EuroSys Conference. ACM, 2018, p. 30.

[25] Pongnumkul, S., C. Siripanpornchana and S. Thajchayapong, "Performance Analysis of Private Blockchain Platforms in Varying Workloads," 2017 26th International Conference on Computer Communication and Networks (ICCCN), 2017, pp. 1–6.

[26] Fan, M. and X. Zhang, "Consortium Blockchain Based Data Aggregation and Regulation Mechanism for Smart Grid." *IEEE Access* 7 (2019): 35929–35940.

[27] Vujičić, D., D. Jagodić and S. Ranđić, "Blockchain Technology, Bitcoin, and Ethereum: A Brief Overview," 2018 17th International Symposium INFOTEH-JAHORINA (INFOTEH), 2018, pp. 1–6.

[28] Dean, J. and S. Ghemawat, LevelDB, https://github.com/google/leveldb

[29] Jansen, M., F. Hdhili, R. Gouiaa and Z. Qasem"Do Smart Contract Languages Need to Be Turing Complete?". In: Prieto J., Das A., Ferretti S., Pinto A., Corchado J. eds.: *Blockchain and Applications. BLOCKCHAIN 2019. Advances in Intelligent Systems and Computing*, vol 1010. Cham, Springer. (2020).

[30] Szabo, N., "Smart Contracts: Building Blocks for Digital Free Markets." *Journal of Trans human Thought* 16 (1996):1–11.

[31] Zheng, Z., S. Xie, H.-N. Dai, W. Chen, X. Chen, J. Weng, M. Imran, "An Overview on Smart Contracts: Challenges, Advances and Platforms." *Future Generation Computer Systems* 105(2020): 475–491.

[32] Bhagavan, S., P. Rao and L. Njilla (2020) "A Primer on Smart Contracts and Blockchains for Smart Cities. In: Augusto J. C. ed. *Handbook of Smart Cities*. Cham, Springer.

[33] https://www.wired.com/2016/06/50-million-hack-just-showed-dao-human/

[34] Zhou, Q., H. Huang, Z. Zheng and J. Bian, "Solutions to Scalability of Blockchain: A Survey." *IEEE Access* 8(2020): 16440–16455.

[35] Bitcoin Cash, 2019, https://www.bitcoincash.org/.

[36] Bip152, 2019, https://github.com/bitcoin/bips/blob/master/bip-0152.mediawiki.

[37] Luu, L., V. Narayanan, C. Zheng, K. Baweja, S. Gilbert and P. Saxena, "A Secure Sharding Protocol for Open Blockchains," Proceeding ACM SIGSAC Conference Computer Communication Security, 2016, pp. 17–30.

[38] Kokoris-Kogias, E., P. Jovanovic, L. Gasser, N. Gailly, E. Syta and B. Ford, "OmniLedger: A Secure Scale-Out Decentralized Ledger via Sharding," Proceedings IEEE Symposium Security Privacy, May 2018, pp. 583–598.

[39] Bentov, I., R. Pass and E. Shi, "Snow White: Provably Secure Proofs of Stake." *IACR Cryptol. ePrint Archive* 2016 (2016): 919.

[40] David, B., P. Gaži, A. Kiayias and A. Russell, "Ouroboros praos: An adaptively-secure semi-synchronous proof-of-stake blockchain," Proceeding Annual International Conference Theory Application Cryptography Techniques, 2018, pp. 66–98.

[41] She, W., Q. Liu, Z. Tian, J. -S. Chen, B. Wang and W. Liu, "Blockchain Trust Model for Malicious Node Detection in Wireless Sensor Networks." *IEEE Access* 7 (2019): 38947–38956.

[42] Shaik, K., *Why Blockchain and IoT are Best Friends, IBM* www.ibm.com/blogs/blockchain/2018/01/why-blockchain-and-iot-are-best-friends/

[43] Tasatanattakool, P. and C. Techapanupreeda, "Blockchain: Challenges and applications," 2018 International Conference on Information Networking (ICOIN), 2018, pp. 473–475.

[44] Delgado-Mohatar, O., J. Fierrez, R. Tolosana and R. Vera-Rodriguez "Blockchain and Biometrics: A First Look into Opportunities and Challenges". In: Prieto J., Das A., Ferretti S., Pinto A., Corchado J. eds.: *Blockchain and Applications. Advances in Intelligent Systems and Computing*, vol 1010, Cham, Springer. (2020).

13

Cryptographic Foundations of Blockchain Technology

C. Tezcan

Middle East Technical University, Ankara, TURKEY

CONTENTS

13.1 Introduction

Many proposed cryptographic payment systems, both credit card and e-cash-based technologies, are proposed in the last 30 years and almost all of them have failed until the introduction of Bitcoin [1], the first modern cryptocurrency, in 2008. Bitcoin combined the blockchain idea with some other technologies and cryptographic primitives to provide an electronic cash that does not require central repository or authority.

Blockchains are digital ledgers comprised of blocks where cryptographic hash functions provide tamper evidence and tamper resistance. Basically, they enable users in a community to record transactions in a shared ledger such that it is not allowed to change a transaction once it is recorded and it is easy to detect if a malicious user modifies a recorded transaction.

Security of blockchains depends heavily on cryptographic algorithms. For instance, security of Bitcoin and similar blockchains depends on the security of the used digital signature algorithms and cryptographic hash functions. And in time, new cryptographic algorithms are added to blockchains and some old algorithms are broken.

DOI: 10.1201/9781003203957-16

First, theoretical [2] and then practical cryptanalysis [3,4] of SHA-1 hash function shows the importance of cryptographic algorithm selection in blockchains. On the other hand, the addition of new cryptographic algorithms might provide better security or new features to a blockchain. For instance, the Taproot upgrade of the Bitcoin that is scheduled for mid-November 2021 is going to add Schnorr signatures [5] to the blockchain. These signatures have better security properties [6] and allow key aggregation, which can shrink multiple signatures to a single signature [7], saving space and transaction fees.

However, development of a large quantum computer would break digital signature algorithms of current blockchains because they rely on the discrete logarithm problem that can be solved on such a quantum computer in polynomial-time [8]. Thus, the performance and the size of the digital signature algorithms competing in the ongoing Post-Quantum Cryptography Competition [9] might be vital for future blockchains.

Blockchains have evolved in time and different cryptographic algorithms like zero-knowledge protocols or lightweight cryptography found their way in this new technology. For instance, although the anonymity in Bitcoin has tried to be achieved by non-cryptographic means, some cryptocurrencies like Zerocoin [10] and Zerocash [11] use special types of zero-knowledge protocols [12] for anonymity.

Blockchains also have Internet of Things applications. However, some IoT devices might have small memory, computational power, or battery. When modern cryptographic standards are run on resource constrained IoT devices, performance and efficiency become an issue. Thus, blockchains on IoT devices might benefit from lightweight cryptographic algorithms. There are ISO/IEC lightweight hash functions, stream ciphers, and block ciphers. Moreover, finalists of NIST's ongoing lightweight authenticated encryption competition were announced in March 2021 [13].

Thus, the security of a blockchain requires understanding of the cryptographic foundations of blockchain technology. Algorithm and parameter selection significantly affect the security, performance, and efficiency of a blockchain. Thus, this chapter investigates the security of cryptographic algorithms that are used in blockchains. We are going to have an emphasis on Bitcoin, since it is the pioneer of this disruptive technology. Moreover, we are going to focus on lightweight cryptography, which is more suitable for constrained IoT devices.

13.3.1 Methodology

Since the use of digital signatures and hash functions in the Bitcoin blockchain, many other areas of cryptography found their way to blockchain technology:

1. **Hash Functions:** In a blockchain, hash functions can be used for address derivation, securing the block header, securing block data, creating unique identifiers, and creating cryptographic puzzles.

2. **Public-Key Cryptography:** Digital signature algorithms are used for signing transactions in a blockchain. Moreover, threshold cryptography can be used for secret sharing.

3. **Post-Quantum Cryptography:** Building a large quantum computer would render most of the public-key algorithms useless, including the digital signature algorithm and its variants. Quantum-safe digital signature algorithms can be used in blockchains in case of such a technological advancement. However, performance and size of such digital signatures can be vital for blockchain technology.

4. **Lightweight Cryptography:** IoT devices have different hardware capabilities and some cryptographic algorithms in a blockchain might be taxing for resource-constrained IoT devices. In order to have better performance or save battery, lightweight alternatives of cryptographic algorithms should be considered in blockchains running on IoT devices.

5. **Zero-Knowledge Protocols:** Full anonymity in blockchains can be achieved via zero-knowledge protocols, which are methods to prove that one knows a value without revealing it.

This chapter is dedicated to the current and possible future use cases of different cryptographic algorithms in a blockchain. For each related cryptographic subfield, we are going to list broken algorithms, secure algorithms, and secure parameter choices for cryptographic algorithms in order to have better performance and security in a blockchain.

In Section 13.2, we summarize the building blocks of the blockchain technology. Section 13.3 analyzes the security of hash functions that is used in blockchains, lists broken algorithms, and compares hash puzzles used in the proof-of-work consensus algorithms. Security of digital signature algorithms and elliptic curves are provided in Section 13.4. Blockchain security against large-scale quantum computers and post-quantum cryptography are discussed in Section 13.5. Anonymity solutions of zero-knowledge protocols are explained in Section 13.6. Lightweight cryptographic standards are provided in Section 13.7. This chapter is concluded in Section 13.8.

13.2 Blockchain Technology

Blockchains are tamper resistant and tamper evident digital ledgers that are distributed amongst the users. They usually do not require a central authority. They enable users to record transactions in a shared ledger in their community. Once a record is written, users are not allowed to delete or update it. Moreover, it can easily be noticed when a record in the blockchain is modified afterwards.

Bitcoin blockchain is the first cryptocurrency that combined many technologies introduced in 2008 and its ledger started on 3 January 2009. Users within the Bitcoin blockchain use digital addresses obtained from cryptographic algorithms and electronic cash related information is attached to these digital addresses. Digital asset transactions between users are digitally signed using public-key digital signature algorithms and the transactions become publicly available in the blockchain for everyone in the network to see. The distributed structure of the Bitcoin blockchain together with cryptographic hash functions make the blockchain tamper resistant and tamper evident. After the invention of blockchain, thousands of other cryptocurrencies are introduced, which generally have different parameters like used cryptographic algorithms, the total number of available coins, or the block time that represents the time that it takes on average for a block to be written to the blockchain.

Although blockchain technology solves the problem of digital currency, it is still unclear what other problems can be solved with this technology. Note that a blockchain might be helpful if we need all of the following conditions:

1. A shared and consistent data store is needed

2. Data is contributed by more than one entity or auditing is required

3. Records are never updated or deleted after they are written

4. Sensitive data will not be stored as plaintext

5. Control of the data store cannot be assigned to a single entity

6. Tamper-proof log of all data store writes is wanted

Blockchain networks can be categorized in two models: *permission-less* and *permissioned*.

Permission-less networks are open to anybody that wants to participate. Without any authorization, anyone can read the data on the blockchain and write to the blockchain. When attempting to publish blocks, a consensus algorithm requires users to maintain or expend resources like solving cryptographic puzzles in order to prevent malicious users or double spending of cryptocurrencies. Such consensus systems generally reward the publisher of the block. Bitcoin and other cryptocurrencies use permission-less blockchain networks.

In permissioned blockchain networks, some authority authorizes entities so that they can publish blocks. Thus, permissioned blockchain networks have limited participation and allow finer-grained controls.

A blockchain is a distributed ledger of blocks that contain cryptographically signed transactions. Each block consists of block data and a block header. A block header contains metadata about the block and a cryptographic hash link to the previous block's header except for the first block which is generally referred to as the genesis block of a blockchain. Each transaction is signed by the blockchain network user or users who are involved in the transaction. Block data contains signed transactions and other related data. The integrity provided by the cryptographic hash functions makes this process tamper evident. Moreover, blockchain becomes tamper resistant as new blocks are added because it becomes harder to tamper older blocks of the blockchain.

A blockchain network generally does not need a trusted intermediary. The trust within a blockchain network is obtained by having a shared ledger that is secured by cryptographic algorithms and distributed amongst the nodes of the blockchain network [14]. When adding a new block, any conflicts amongst the nodes are resolved automatically based on pre-determined rules.

Unlike traditional databases, blockchain technology uses ledgers that are append only. Thus, a written value is never deleted or changed but a transaction affecting that value is appended to the ledger, providing full transactional history that results in transparency. Distributed ownership of ledgers has advantages over centralized ownership of ledgers. The centrally owned systems might be insecure because a weakness in the network or computer systems might leak personal information. Moreover, it is assumed that the best security measures are taken and security patches are installed regularly.

A disadvantage of distributed ledgers are conflicts. Different blocks might be published around the same time and some nodes in the network might have different versions of the blockchain. Conflicts should be resolved for the consistency of the blockchain and generally they are solved in a short time. In a conflict, the blockchain that becomes longer is generally considered as the correct version. These conflicts are sometimes referred to as forks, but forks are different than conflicts and explained in Section 13.2.3. Block structure is explained in Section 13.2.1 and consensus models are summarized in Section 13.2.2.

13.2.1 Blocks

Transactions are stored in blocks in a blockchain. In some blockchains, a block has to be published in a pre-determined time for security. For instance, a block is published in the Bitcoin blockchain in 10 minutes and in the Ethereum network in 1 minute on average. For this reason, a block can have no transaction at all, which happens many times in cryptocurrencies.

The block size in a block chain generally has an upper bound. For instance, a Bitcoin network has 1 MB block size limit and after a soft fork this size was virtually increased to 4 MB via the SegWit upgrade. For this reason, number of transactions per second or total number of transactions in a block might be upper bounded in a blockchain.

Data stored in a blockchain depends on the blockchain implementation. Generally, data sent to a blockchain network contains a user's address, user's public key, transaction inputs and outputs, and a digital signature.

Although blockchains are primarily used for transferring digital assets, it can be used to permanently write data on the blockchain and publicly announce it. Some blockchains might support smart contracts in which transactions can be used to process the data that is being sent and store some of the results on the blockchain.

A transaction submitted by a user to the blockchain becomes a candidate transaction. When a publishing node includes these transactions in a block, they are written to the blockchain. Amongst the submitted transactions, some or all of the validated and authentic transactions are stored as the block data. The metadata of the block is stored in the block header.

Data fields of block headers and block data differ significantly amongst blockchains. Generally, a block header contains fields like block height, hash of the current block data, hash of the previous block header, timestamp, size of the block, and nonce. Nonce is a random value generally used in blockchain networks that requires mining in order to make the hash value less than a pre-determined value called the block difficulty. A set of transactions included within the block and some other related data about the block may appear as the block data.

13.2.2 Consensus Models

The whole network agrees on the initial block that is also referred to as the genesis block. The distributed nature of the blockchains introduces a synchronization problem among the nodes when new blocks are added. To avoid conflicts and double spending problem of cryptocurrencies, a consensus model is used. New blocks are added to the blockchain according to the consensus model and each block is linked to the previous block because each block header contains the previous block header's hash value. The most popular consensus models are proof of stake and proof of work for permission-less models. Proof of authority and round robin are common consensus models that are used in permissioned models.

A proof-of-work consensus model is used in permission-less blockchain networks. A cryptographic puzzle that is generally based on hash functions forces the publishing node to hash the block header with different nonces so that the hash output is smaller than the difficulty value of the puzzle. This is an open consensus algorithm and generally the solver of the cryptographic puzzle is rewarded. This is generally how new coins are introduced in cryptocurrencies. This is referred to as mining. The resources used for cryptocurrency mining increased so much in the recent years so that some cryptocurrencies like Ethereum plan to switch to proof-of-stake consensus model in the future. For instance, as of 2021, the yearly electricity consumption of Bitcoin mining is estimated to exceed 140 terawatt-hours, which is higher than the yearly electricity consumption of some of the European countries like Norway, which consumes more than 120 terawatt-hour yearly.

A proof-of-stake consensus model is used in permission-less blockchain networks. It is introduced as an energy-saving alternative for proof of work. The idea is that users will want the system to function properly as the stake invested by them increases. Publishing a new block is determined by the amount of stakes spent by the users. How the stakes are used may vary from network to network.

In a round-robin consensus model, nodes take turns in creating blocks. When a publishing node is unavailable on its turn to publish a block, a time limit may be enabled to prevent system halt. This model does not require any cryptographic puzzles, has very low power requirement, and prevents any node from creating a majority of the blocks.

A proof-of-authority consensus model relies on the high level of trust amongst the publishing nodes since they know their real-world identities. It relies on the fact that misbehavior of a publishing node would affect their reputation.

A proof-of-elapsed-time consensus model requires secure hardware that can generate a random number that determines the wait time of a publishing node.

13.2.3 Forking

Some intentional changes to a blockchain data structure or network's protocol might be necessary for many reasons. Such changes are called forks and can be categorized as soft and hard forks.

A soft fork is a backward compatible change to a blockchain implementation. Non-updated nodes may not understand valid transactions that come from updated nodes but they can still continue to transact with them. Increasing the data block size limit virtually to 4 MB via the SegWit upgrade from 1 MB in the Bitcoin network is a soft fork example. Adding a new cryptographic algorithm to a blockchain requires a soft fork. For instance, Bitcoin's Taproot upgrade, which is scheduled for November 2021, is going to create a soft fork by adding Schnorr signatures [5] to Bitcoin.

A change to a blockchain implementation that is not backwards compatible is referred to as a hard fork. These hard forks sometimes might be unintentional and might be caused by a software error. Intentional hard forks might be due to a flaw found in the cryptographic algorithms used in the blockchain. When a hard fork is introduced to a cryptocurrency, this actually generates two cryptocurrencies where the users' data is duplicated in both blockchains. For example, in 2016, a hacker stole around $50 million US dollars' worth of tokens from a decentralized autonomous organization and Ethereum reversed this theft. However, the original blockchain continued as Ethereum Classic. Thus, the hard fork resulted in two different cryptocurrencies.

13.3 Cryptographic Hash Functions

Definition 1: (Cryptographic Hash Function): A hash function is a deterministic algorithm $h(.)$ that has a fixed length output, which is referred to as *message digest*, while the input length is variable. For cryptographic hash functions, the following three properties are required for security:

1. **Preimage Resistance:** When the output y is known, finding an input x such that $h(x) = y$ should be infeasible.

2. **Second Preimage Resistance:** When an input x_1 and the output y is known where $h(x_1) = y$, finding x_2 such that $h(x_2) = y$, where $x_1 \neq x_2$, should be infeasible.

3. **Collision Resistance:** Without any condition on the output y, finding any different x_1 and x_2 such that $h(x_1) = y$ and $h(x_2) = y$ should be infeasible.

Although it is not included in the definition, we desire message digest to appear random. This property is necessary for hash puzzles. For an n-bit message digest size, a brute force collision attack requires around $2^{n/2}$ hash calculations due to the birthday paradox. However, brute force attacks for finding a preimage or a second preimage requires around 2^n hash calculations. Although collision resistance and second preimage resistance look similar, an input is fixed in second preimage resistance which makes it harder for the attacker. For most of the blockchain applications we require second preimage resistance. However, using a hash function that is not collision resistant is not suggested since it reduces the trust in the design. Moreover, attacks always get better and a collision attack might lead to a second preimage attack in the future.

Many hash functions like MD4, MD5, RIPEMD, SHA-2, and SHA-1 use Merkle-Damgard construction. Merkle-Damgard construction converts a collision-resistant one-way function into a collision-resistant hash function. As shown in Figure 13.1, firstly a padding is applied to the input M in Merkle-Damgard construction and then it is divided into t blocks M_i. At the i-th step, the compression function f produces h_i from the inputs h_{i-1} and M_i as $h_i = f(h_{i-1}, M_i)$. An initialization vector (IV) is used as h_0. A finalization function is generally applied to the final output h_t to produce message digest.

It was shown that MD4 and MD5, which are designed by Rivest, are not collision resistant and many collisions can be produced in milliseconds. However, these designs influenced many later hash functions like SHA-2, RIPEMD, and SHA-1.

SHA-1 was developed by NSA in 1993 and a small modification was made in 1995 in which a single bitwise rotation is added. The initial version was generally referred to as SHA-0 and practical collisions can be found for SHA-0. SHA-1 has a block size of 512 bits and a message digest size of 160 bits. It has 80 rounds and its type is *Merkle-Damgard*.

Although a theoretical collision attack on SHA-1 was provided in [2] as early as 2004, it was only in 2017 that a practical collision was obtained [4]. Note that a collision attack on a hash function only affects scenarios where collision resistance is required. However, recently, chosen-prefix collisions on SHA-1 is provided in [3]. This new attack is much more threatening for real protocols because it allows building two messages with arbitrary prefixes that have the same hash values. As an example, the authors of [3] provided a PGP/GnuPG impersonation attack. Therefore, SHA-1 should be avoided at all costs.

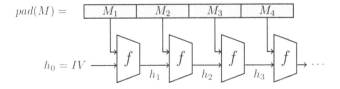

FIGURE 13.1

Merkle-Damgard construction. Figure is drawn by Jeremy Jean and it is publicly available [15] under Creative Commons license CC0.

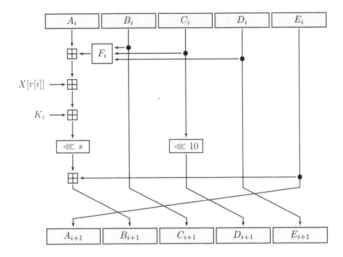

FIGURE 13.2
i-th step of RIPEMD-160 hash function. Figure is drawn by Jeremy Jean and it is publicly available [15] under Creative Commons license CC0.

RIPEMD-160 hash function [16] is also a Merkle-Damgard construction. It has a block size of 512 bits and consists of 80 steps. One step of RIPEMD-160 is provided in Figure 13.2 where $K[i]$ represents round constants and $X[r[i]]$ represents a 32-bit part of the message block. Having 160-bit message digest size like SHA-1 and not being broken makes RIPEMD-160 a good candidate for applications where small hash sizes are required. Moreover, unlike SHA-1, RIPEMD-160 was designed by academicians.

SHA-2 was developed by NSA and first published in 2001. Its block size is either 512 bits or 1,024 bits, depending on the message digest. The message digest length can be 224, 256, 384, or 512 bits. Generally, this number is used with the algorithm name. For example, SHA-256 is SHA-2 with 256-bit message digest length. The number of rounds is either 64 or 80, which also depends on the message digest size. It is also a Merkle-Damgard-type hash function. The design is similar to its predecessors SHA-1, MD5, and MD4.

Although SHA-2 is not broken, its predecessors SHA-1, MD5, and MD4 are broken. For this reason, NIST (National Institute of Standards and Technology) announced a SHA-3 Competition that took place between 2007 and 2012. This competition received 64 submissions and the winner KECCAK became SHA-3. Its message digest sizes are the same as SHA-2 to allow transition from SHA-2 to SHA-3 if necessary. The block sizes are 1,152, 1,088, 832, and 576 bits corresponding to 224-, 256-, 384-, and 512-bit message digest sizes. The number of rounds is 64 or 80. Unlike SHA-2, SHA-3 is a sponge construction. Sponge functions have relatively larger internal states and a permutation modifies this internal state at each step. A small part of the state is used for producing output.

It is a common mistake to indicate that a hash function input can be anything and therefore its input space is infinite because standards generally limit the input size by 2^{64} bits. However, such a limit allows 2 to the power of 2^{64} different input, which is virtually infinite.

Cryptographic hash functions have many different use cases in a blockchain network such as address derivation, securing the block header, securing block data, creating unique identifiers, and creating cryptographic puzzles.

Currently, the Bitcoin script allows the use of three hash functions: SHA-1, RIPEMD-160, and SHA-256. These are provided by the opcodes OP_SHA1, OP_RIPEMD160, and OP_SHA256, respectively. However, Bitcoin transactions mostly use OP_HASH160 and

OP_HASH256 opcodes. OP_HASH256 applies SHA-256 twice and OP_HASH160 applies SHA-256 and RIPEMD-160 successively.

Note that the output of OP_HASH160 is 20 bytes and the output of OP_HASH256 is 32 bytes. Saving 12 bytes by using OP_HASH160 instead of OP_HASH256 is useful for having shorter addresses and using smaller space on the blockchain, which would also reduce the transaction fee.

Blocks in a blockchain are identified by block headers. Headers generally contain metadata about the block and the hash of the transactions included in that block which provides integrity without the use of signatures. For example, a Bitcoin block header contains a 4-byte version number, 32-byte Merkle root, 32-byte previous block hash, 4-byte difficulty target of the cryptographic mining puzzle, 4-byte timestamp, and the 4-byte nonce that is used by the miner that mined that block.

Since the block header contains the hash of the block data, an attacker that is capable of computing second preimages might gather different set of transactions that would lead to the same hash and claim that those transactions are the correct ones. Moreover, such a strong adversary can focus on a single transaction, instead of the whole block. Forging a transaction that hashes to a transaction hash value that appeared in a previous block can cause confusion amongst the nodes.

Blockchains like Bitcoin use Merkle Trees as a data structure for fast verification of the transactions in a block. In a Merkle Tree, hash of the transaction are the leaf-level values of the tree and pairs of hashes are hashed together to obtain parents going upwards until the root hash is obtained. If an attacker forges a transaction that has the same hash with another transaction that is recorded in the block and therefore in the Merkle Tree, a full blockchain node that keeps every transaction can realize that the forged transaction is a second preimage of a transaction they recorded. However, Simplified Payment Verification (SPV) nodes do not store every transaction and they may verify the forged transaction since its hash eventually reaches to the Merkle root. Therefore, it is vital to have a second preimage resistant hash function in a blockchain to secure the transactions and the blocks.

In a blockchain, users sign transactions by their private key and their public key works as their identity. To shorten the size of these identities, their hashes are used most of the time. Therefore, once a transaction needs to be verified, first a node checks if the provided public key hashes to the recorded hash of the identity. An adversary that is capable of generating second preimages can claim that a transaction output that belongs to a public key pk_1 actually belongs to the public key pk_2 because $h(pk_1) = h(pk_2)$. However, this is not a problem for a blockchain's security because the attacker that finds such a pk_2 would not be able to obtain the private key that corresponds to pk_2. Thus, the adversary would not be able to sign transactions by impersonating pk_1. For example, an adversary cannot use such a second preimage to transfer Bitcoin or cryptocurrency of others.

Since it is vital to keep private key secret, many cryptocurrency wallets use password-based key derivation functions (PBKDF) where the user-selected password is hashed many times to obtain a secret key for symmetric key encryption algorithm like AES. This key is used to encrypt the private key. The users must choose strong passwords and provide them whenever they want to make a transaction. Thus, hash functions used in PBKDFs must be preimage resistant.

13.3.1 Hash Puzzles

Hash functions are also extensively used in proof-of-work consensus algorithms. Many cryptocurrencies use power of work to reach a consensus which requires computation of hash functions many times. Thus, miners with higher hashrates earn more crypto-currencies. In order to prevent designing custom ASICs for mining, many crypto-currencies use memory-intense algorithms like scrypt, X11, or X16. However, there are many cases where ASICs are built for algorithms that were initially considered ASIC resistant. In some cases, they are used to perform 51% attack on the related crypto-currencies. The expected time required to add a new block in permission-less blockchains where any node can propose a new block is generally pre-determined in order to avoid problems that can be introduced by the latency in the network. In cryptocurrencies, this is generally determined by the block difficulty and it is updated according to the network's block addition speed. Currently in Bitcoin, blockchain block difficulty is chosen so that a block is expected to be added to the blockchain every 10 minutes. Bitcoin's difficulty is recalculated in every 2,016 blocks, which happens around every 2 weeks. In Ethereum blockchain, the difficulty is chosen so that a new block is added in 1 minute. Note that a change in the expected time for block addition would create a hard fork in the blockchain.

Bitcoin requires SHA-256 to be applied twice to a block to get a hash value that is smaller than a value determined by the current difficulty level. Thus, miners aim to obtain a hash value starting with a lot of zero bits. In order to succeed, miners can change a 32-bit nonce value and fields in the coinbase transaction which is used to award the miner for solving this hash puzzle. For example, a coinbase transaction can use the 256-bit previous transaction hash field as a nonce since a coinbase transaction has no previous transaction. Although finding a solution to the hash puzzle is hard, verifying a solution is fast which is a must for the blockchain to function properly.

Initially, Bitcoin miners used CPUs and then moved on to GPUs. However, today it is almost impossible to mine Bitcoin using CPUs, GPUs, or FPGAs because ASICs out-perform these devices in Bitcoin mining. This is somewhat against the decentralization idea behind the Bitcoin because only the nodes with special devices now have the control of the Bitcoin network. For this reason, many cryptocurrencies tried to obtain hash puzzles that are ASIC-resistant. Initial designs like scrypt [17] focused on puzzles that require a lot of memory which was assumed to slow down ASICs. Scrypt requires the miner to store successive hash values in the memory. Litecoin is an example for a cryptocurrency that uses scrypt and it aimed for ASIC-resistance. However, many ASICs for scrypt were produced in the past years that outperform GPU miners. The memory requirement of scrypt algorithm can be increased for better ASIC-resistance but this would slow down the verification of the hash puzzle solution, rendering the blockchain unusable.

Another approach used in practice is to use many different hash functions successively instead of a single hash function like Bitcoin's SHA-256 puzzle. Without a theory, this approach was assumed to be ASIC-resistant. In this respect, X11 algorithm is used by many cryptocurrencies which is successive use of 11 hash algorithms which were in the SHA-3 competition. Similarly, X16 algorithm uses 16 different hash functions making it harder for ASIC designers to design efficient hardware.

Having a hash function that is ASIC resistant is also necessary for password hashing. Because slowing down password hashing also slows down brute force password breaking attacks. Many previous password hashing solutions are ad hoc methods and there were no dedicated designs for password hashing. For this reason, a Password Hashing Competition (PHC) was held that ran from 2013 to 2015. This competition re-ceived 24 candidates and the winner was *Argon2*. However, some other candidates like *Lyra2* and its variants received attention from cryptocurrency designers and they are used in hash puzzles of some cryptocurrencies.

In its proof-of-work algorithm *ethash*, Ethereum uses Keccak but it should be noted that it is slightly different than the standardized version of Keccak, namely SHA-3. A late change in the standardization process of SHA-3 resulted in a different padding than the one used in ethash. Thus, sometimes the used hash algorithms in ethash are referred to as Keccak-256 and Keccak-512, instead of SHA3-256 and SHA3-512.

The difference between the proof-of-work algorithms of Ethereum and Bitcoin is not only the used hash algorithm. Ethash is based on Dagger-Hashimoto in order to be a memory-hard mining algorithm. Ethash generates a 16-MB pseudorandom cache from a seed that can be obtained from the previous block headers. Light nodes only store this cache. However, full nodes and miners generate 1-GB data set from the cache. The dataset size increases linearly with time. This dataset is actually a directed acyclic graph (DAG) which is created every epoch (currently epoch increases every 30,000 blocks). At the beginning of 2021, ethash DAG size exceeded 4 GB, preventing older GPUs and ASICs to continue mining.

In all of the hash puzzles we considered so far, the aim of the miners is to obtain a hash value that starts with many zero bits. Therefore, hash functions used in the hash puzzle must be preimage resistant. Since only some of the first bits of the hash value are fixed to zero, we actually need a hash function that is near-preimage resistant (i.e. only some bits of the hash output are fixed). Therefore, an adversary that has a fast method to obtain small hash values can dominate the blockchain. Solving the hash puzzles faster than the rest of the network would give the adversary the complete control of the blockchain. So far no near-preimage attacks are discovered for the hash functions that are standardized or used in hash puzzles.

13.4 Public-Key Cryptography

In public-key encryption and digital signature algorithms, users generate private and public key pairs. In a blockchain, public-key cryptography provides authentication via digital signatures. It can be used to verify that the person signing the transaction that transfers a value to another person has the private key. Thus, transactions are digitally signed by private keys and signatures can be verified by the public keys. Moreover, public keys are used as identities.

Public-key cryptography provides different algorithms for encryption, key exchange, and digital signatures. The security of these algorithms depend on the intractability of some mathematical problems like discrete logarithm or integer factorization:

Definition 2: (Integer Factorization Problem): Let p, q in $\mathbb{Z}^+$. Given $n = pq$, finding q or p is called the IFP (integer factorization problem).

In cryptography, most of the time q and p are chosen to be prime numbers. Note that it is easy to compute n when one knows p and q but known algorithms require much more time to find prime factors of n. Security of the RSA public-key encryption algorithm depends on the intractability of this problem. The best-known factorization algorithms show that in order to have 128-bit security, one needs 3,072-bit RSA modulus n.

Definition 3: (Group): A group G in algebra is a set of elements together with a binary operation on G that combines two elements in G and form a third element in G satisfying associativity, identity element, and inverse element group axioms. i.e. In a group $a\,(b\,c) = (a\,b)\,c$, there is an identity element, and every element has an inverse with respect to the group operation. An element g in G is called a generator if every element of G can be obtained from the successive application of the group operation on g. In additive groups, a many addition of g is represented as $b = ag$. Similarly, in multiplicative groups a times multiplication of g is represented as $b = g^a$.

Definition 4: (Discrete Logarithm Problem): Let g be a generator of a group G. Given b and g such that $b = g^a$, finding a is called the discrete logarithm problem (DLP).

Note that the intractability of DLP depends on the group G. If our group consists of elements in $\mathbb{Z}_p = \{0, 1, 2, ..., p - 1\}$ where p is a prime number and our group operation is multiplication, it is easy to compute $g^a(mod\ p)$ but it is harder to find a when we only know b, g, and p.

This problem becomes even harder if we use the points of an elliptic curve as the elements of a group G. This is known as ECDLP (Elliptic Curve Discrete Logarithm Problem). The best-known algorithms on ECDLP show that in order to have 128-bit security, one needs 256-bit discrete logarithm key and 3,072-bit group.

An elliptic curve has infinitely many points when it is considered in $\mathbb{R}$ or $\mathbb{C}$. Thus, we consider elliptic curves over finite fields. For an odd prime $p > 3$, an elliptic curve over F_p is defined by $y^2 = x^3 + ax + b$ and $4a^3 + 27b^2 \neq 0\ (mod\ p)$ where $a, b \in F_p$. The set $E(F_p)$ consist of all points (x, y) where $x, y \in F_p$ satisfying the elliptic curve equation and a special point O called the point at infinity that functions as an additive identity element. A similar definition is valid for elliptic curves that are defined over finite fields F_q where q is not a prime number but it is a power of 2, i.e. $q = 2^m$.

The addition of two points Q and P on the elliptic curve provides the output point R where addition is defined by drawing the line that passes through P and Q and taking the negative of the y-coordinate of the third point where this line intersects the elliptic curve.

Definition 5: (Elliptic Curve Discrete Logarithm Problem): Let E be an elliptic curve over a finite field Fq. Let N be the order of the group $E(Fq)$ and let $P \in E(Fq)$. Given $j = ord(P)$ and $Q \in <P>$, finding the unique integer $m \in [0, 1, 2, ..., j - 1]$ such that $Q = mP$ is called the elliptic curve discrete logarithm problem (ECDLP).

Digital signature algorithm (DSA) and its elliptic curve analogue ECDSA (elliptic curve digital signature algorithm) [18] are used in a blockchain mostly for signing transactions.

DSA Parameter Generation:

1. Choose a good hash function $H(.)$ (e.g. SHA-2, SHA-3)
2. Choose key lengths L and N (e.g. (3072,256))
3. Choose N-bit prime q
4. Choose L-bit prime modulus p such that $p - 1$ is a multiple of q
5. Choose g (its multiplicative group order modulo p must be q)

DSA User Key Generation:

1. Randomly choose x where $0 < x < q$
2. Calculate $y = g^x mod\ p$

   x is the secret and y is the public key.

DSA Signing:

1. Randomly choose k per message m where $0 < k < q$
2. Calculate $s = k^{-1}(H(m) + xr) \bmod q$ where $r = (g^k \bmod p) \bmod q$

    The signature is the pair (r, s).

DSA Verification:

1. Calculate $v = (g^{u_1} y^{u_2} \bmod p) \bmod q$ where $u_2 = rw \bmod q$, $u_1 = H(m)w \bmod q$, and $w = s^{-1} \bmod q$

    If $r = v$, then the signature is verified.

The security of the digital signature algorithm also depends on DLP. Unlike the ordinary DLP and IFP, no subexponential-time algorithm is found so far for ECDLP. Thus, algorithms using elliptic curves have greater strength-per-key-bit and therefore signatures using elliptic curves result in shorter user keys. Note that the Bitcoin uses ECDSA.

Known attacks on the ECDLP force us to carefully select the elliptic curve parameters:

1. To avoid the Pohlig–Hellman [19] attacks and Pollard's Rho [20,21] attacks, j should be a large prime of at least 224 bits because they have a time complexity of $O(\sqrt{j})$.
2. The MOV [22] and Frey–Ruck [23] attacks reduce the ECDLP on $E(Fq)$ to the DLP in F_t, where $t = q^l$ for some integer l, by using a Weil pairing on $E[j]$. When $gcd(j, q) = 1$, the integer l is the smallest value satisfying $q^l \equiv 1 \pmod{n}$. This reduction is polynomial in terms of the number of operations in F_t. Therefore, to avoid the MOV attack, j should not divide $q^l - 1$ for each $1 \le l \le C$ where $C = 20$ is sufficient. This required property of j cannot be obtained when the curve is supersingular.
3. When q is prime, the elliptic curves whose trace of Frobenius t is 1 and $q = N$ are called anomalous curves. These curves resist the MOV attack. However, using q-adic elliptic logarithm [24,25] or considering the q-primary part of the subgroup generated by P [26], one can give a linear time method to solve the ECDLP. To avoid these attacks, N should be different from q.

Known attack techniques like GHS's Weil Descent [27] can break ECDLP on some elliptic curves faster than Pollard's Rho method but it was shown in [28] that this attack is ineffective against Koblitz curves which are elliptic curves defined over F_2 and for elliptic curves against NIST's recommended fields.

Therefore, in practice, we need ordinary, non-anomalous (i.e. $N \ne q$) elliptic curves with group orders divisible by a large prime number of at least 224 bits. In practice, an elliptic curve is selected randomly and kept if it satisfies these conditions. Thus, determining the order of the group $E(F_q)$, which is also known as the point counting problem, is of critical importance in elliptic curve cryptography. Schoof's point counting algorithm [29,30] solves this problem efficiently. Although working with random curves is not as efficient as working with special curves like Koblitz curves [31], it is practical. Therefore, it is a natural question to ask if the ECDLP is equally hard for every elliptic curve over the same finite field having the same number of points. This problem is answered in terms of random self-reducibility.

Bitcoin uses the elliptic curve $y^2 = x^3 + 7$ over F_p where $p = 2^{256} - 2^{32} - 2^9 - 2^8 - 2^7 - 2^6 - 2^4 - 1$ for ECDSA. This curve is called *secp256k1* and the same curve with the same public/private key pairs can be used in some other algorithms like Schnorr signatures [5] which also relies on the hardness of DLP. Currently Bitcoin only supports ECDSA. However, Taproot fork that is expected to happen in November 2021, will also provide support for Schnorr signatures.

Unlike DSA, ECDSA allows us to choose many elliptic curves with the same number of points on them. Thus, a natural question arises: Does the DLP equally hard for elliptic curves that have the same number of points? This question is older than Bitcoin because in cryptography, a common practice is to use random curves. Thus, cryptographers wondered if it is possible to come across a weak elliptic curve. In the case of Bitcoin, we wonder if *secp256k1* is a weak curve for which DLP is easier to solve. In [32], it is shown that almost all elliptic curves over the same finite field having the same number of points have the same difficulty of discrete logarithm by showing polynomial time random self-reducibility of the ECDLP amongst these curves. Orders in imaginary quadratic fields play an important role in this result because the endomorphism ring of an ordinary curve is isomorphic to an order in an imaginary quadratic field. For a fixed N and m, elliptic curves that are isomorphic to the same order in an imaginary quadratic field form the vertices of an isogeny graph and prime degree isogenies between them of degree less than m form the edges. The polynomial time random self-reducibility of the ECDLP is achieved by navigating this isogeny graph i.e. If there are polynomial amount of weak elliptic curves for which the ECDLP is easy, then the ECDLP for the rest of the elliptic curves over the same finite field having the same number of points can be solved by transferring the ECDLP for a strong curve to ECDLP for a weak curve by computing polynomial amount of random isogenies between elliptic curves. Thus, it is reasonable to assume that Bitcoin's elliptic curve *secp256k1* is as secure as any elliptic curve over the same finite field having the same number of points. Moreover, NIST announced that they are going to include more elliptic curves to their portfolio [33]. Thus, NIST is expected to include Bitcoin's elliptic curve *secp256k1* for interoperability.

Although Schnorr Signatures were invented in 1990, their patent expired in 2008. Bitcoin was also launched in 2008 but at the time Schnorr Signatures were not as popular as ECDSA, which was widely used and tested. This is why Bitcoin did not support Schnorr Signatures at the beginning and now a soft fork is required to add a new cryptographic algorithm. This soft fork upgrade for Bitcoin called Taproot is scheduled for November 2021.

The main advantage of Schnorr Signatures against ECDSA or DSA is key/signature aggregation. Bitcoin and most of the blockchains allow multi-signatures, which provides transactions that are signed by multiple parties. A *multisig* transaction that is signed by n parties contains n public keys which increases the transaction size and the computation required for the verification of the transaction. However, Schnorr public keys and signatures can be aggregated [7] so that n parties that are going to sign a transaction can combine their public keys to form a single public key, without a need to trust each other. Then they can sign the same message with their private keys. Combining these n signatures forms a single signature which can be verified by the aggregate public key. Thus, a verifier must only verify a single signature and a public key to check that all of the n parties signed the message. Therefore, it is expected to have smaller transaction sizes resulting in fee savings. This would provide extra privacy for Bitcoin users because it makes it impossible to distinguish single signature spends from *multisig* spends. Note that *multisig* transactions are also used by a single party when the transaction requires the transferring assets from more than one previous transaction.

Moreover, it was shown in [6] that unlike DSA signatures, Schnorr signatures do not rely on collision-resistant hash functions which allows 25% smaller signatures with the same amount of security.

Randomness plays a very important role in DSA, ECDSA, and Schnorr Signature Algorithm. We require random numbers for key generation and every signature. Thus, a strong random number generator (RNG) is a must for security of these algorithms. Using a weak random number generator might result in predictable secret keys during the key generation step. Moreover, we need random numbers for every signature and these numbers should be secret and different. Because, if the attacker knows the random number k that is randomly selected during the signing step, they can easily extract the secret key x from the signature (r, s).

Random numbers generated for signing must also be distinct. Using the same k for signing two different transactions allows an attacker to capture this k and consequently obtain the secret key x. In the past, some Bitcoin users used the same k for signing two different transactions due to a Java bug in the Java SecureRandom class on Android and attackers stole their Bitcoins by obtaining secret keys from these transactions.

Many devices like most of the modern IoT systems-on-a-chip (SoCs) have a dedicated hardware RNG peripheral. However, incorrectly calling the hardware RNG might result in non-random numbers because sometimes developers fail to check error code responses. When the hardware RNG runs out of entropy, the produced random numbers might have low entropy. Thus, using the entropy generated from the hardware RNG as a random number might result in weak numbers. Using CSPRNG (cryptographically secure pseudorandom number generator) on top of the entropy source of IoT devices can prevent this kind of problems.

One way to keep a private key secret is to use PBKDF. Another cryptographic way to secure the private key is to use threshold cryptosystems like Shamir's Secret Sharing Scheme [34]. We call it a *(t,n)*-threshold scheme when any t out of n parties can combine their shares to obtain a secret. This way a user can keep their shares in different devices so that it stays secure when the number of compromised devices is less than t. This scheme uses the polynomial interpolation idea which allows to construct a t-1 degree polynomial when t points on it are known. Although this looks like a multi-signature scheme where $t = n$, in multi-signature scheme we have n different public/private key pairs. Here, n shares of a single secret are distributed among n parties.

13.5 Post-Quantum Cryptography

Development of a large quantum computer would solve ECDLP, DLP, and IFP on a quantum computer in polynomial-time [8]. Public key cryptography algorithms whose security depend on the intractability of these problems would easily be broken if such a quantum computer is ever built. This includes leakage of every blockchain user secret key that uses DSA, ECDSA, or Schnorr signature. Thus, we need quantum safe algorithms for public key cryptography. For this reason, NIST is currently organizing a Post-Quantum Cryptography Competition.

On 24 February 2016, NIST called for post-quantum algorithm submissions and NIST's internal report 8105 on Post-Quantum Cryptography was published in April 2016. The deadline for submissions was 30 November 2017 and the competition received 69 complete and proper submissions. NIST organized the First Post-Quantum Cryptography Conference in April 2018 and on 30 January 2019 NIST announced 26 algorithms as the Round 2 candidates. A second Post-Quantum Cryptography Conference was held in August 2019. Recently, NIST announced the Round 3 Finalists. For digital signature algorithms, the finalists are *CRYSTALS-DILITHIUM, FALCON*, and *Rainbow*. The finalists of key-establishment and public-key encryption are *SABER, NTRU, CRYTALS-KYBER*, and *Classic McEliece*.

The competition is expected to be finished before 2024 and draft standards should be available around that time. Blockchain designers should consider including these quantum-safe digital signature algorithms in their implementations or future upgrades. Recall that adding new cryptographic algorithms to a blockchain requires soft forks.

13.6 Zero-Knowledge Protocols

Cryptocurrencies like Bitcoin provide pseudo-anonymity. That is, anybody can see every transaction in the blockchain but nobody know which account belongs to whom in real life. In order to obtain anonymity, users use some non-cryptographic techniques like mixing in which n people get together and sign a multi-signature transaction to hide the receiver and the sender of a single transaction. Such techniques are not always practical and it requires trust amongst the participants. Two protocols, Zerocoin [10] and Zerocash [11], incorporate anonymity at the protocol level by using zero-knowledge protocols. The main idea behind zero-knowledge protocols is that one party can prove that they know a value x without revealing it.

Zerocoin protocol uses two coins A and B to obtain anonymity. A user obtains anonymity by destroying their A coins and minting same amount of B coins. Although this provides anonymity, transaction amounts are still visible in Zerocoin. One of the cryptographic building blocks of Zerocoin protocol is the use of RSA accumulators which requires a one-time trusted setup. A trusted party is needed for the choice of two primes q and p to construct $N = pq$ and then destroy q and p. Thus, the security depends on the IFP. Anybody that can factorize N can create new coins without being detected. Zcoin was the first cryptocurrency that used this protocol but then moved on the Sigma protocol [35], which removes the trusted setup feature of the Zerocoin protocol.

To provide full anonymity, Zerocash uses zk-SNARKs [12]. zk-SNARKs make zero-knowledge proofs much more compact and efficient to verify. This efficiency removes the necessity of having a basecoin that is used in Zerocoin. Moreover, public ledger no longer contains transaction amounts in Zerocash. However, over 1 GB of random/secret public parameters are needed to set up Zerocash compared to a single public number N that is used in Zerocoin. How else zero-knowledge proofs or protocols can be used in blockchain technology remains an open problem.

13.7 Lightweight Cryptography

The number of resource-constrained devices has increased significantly and having cryptographic algorithms with the following properties became a necessity:

- require less energy
- require less power
- have small latency
- provide better throughput
- have side-channel resistance

Lightweight cryptography focuses on resource-constrained devices and tries to provide solutions that are tailored for them. There are ISO/IEC standards for lightweight cryptographic algorithms. Moreover, currently NIST is organizing a Lightweight Cryptography Competition to standardize one or more algorithms.

In symmetric cryptography, the keys used for encryption and decryption are the same or closely related. These encryption algorithms are mainly categorized as stream ciphers or block ciphers. Almost all permissionless blockchains do not encrypt the blocks because the data is available for everyone. On the other hand, permissioned blockchains may prefer encryption of the blocks. However, this means that additional measures must be taken in order to generate, distribute, and protect the secret key or keys.

Advanced Encryption Standard (AES) withstands all cryptanalytic attacks after more than 20 years of investigation. Moreover, modern CPUs come with hardware level AES instruction sets (e.g., Intel's AES-NI). Therefore, AES is arguably the first choice for the encryption of blocks in a blockchain. However, if the blockchain is going to be stored also in constrained devices, then using a lightweight cipher might be a better solution. There are three ISO/IEC standards for lightweight block ciphers: *PRESENT*, *CLEFIA*, and *LEA*.

PRESENT was designed in 2009 by Bogdanov *et al.* [36]. It is a hardware-oriented design focusing on a low hardware footprint. PRESENT is an SPN design. Its block size is 64 bits and supports 128-bit and 80-bit keys. PRESENT's two rounds are provided in Figure 13.3 and it consists of 31 rounds.

CLEFIA was designed by Shirai *et al.* [37] and it is a software oriented algorithm. CLEFIA is a Feistel network. Its block size is 128 bits. It supports keys of size 128, 192, and 256 bits. The number of rounds for these key lengths are 18, 22, and 26, respectively. Details of CLEFIA are provided in Figure 13.4.

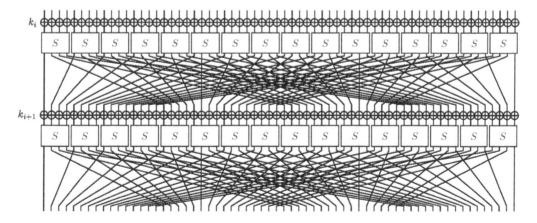

FIGURE 13.3
Two rounds of PRESENT cipher. Figure is drawn by Jeremy Jean and it is publicly available [15] under Creative Commons license CC0.

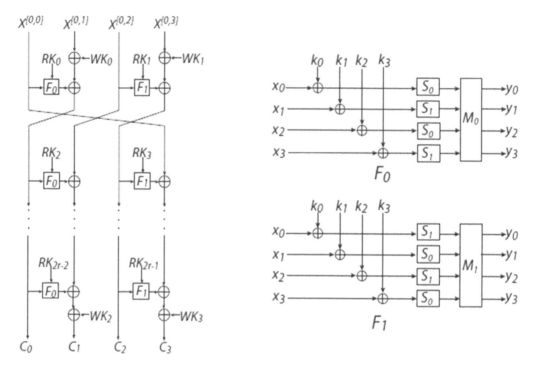

FIGURE 13.4
Block cipher CLEFIA (on the left) and its F_0 and F_1 functions. Reprinted by permission from Springer [39] and reprinted from [40] with permission from Elsevier.

LEA was designed by Hong et al. [38] and it was added to ISO/IEC lightweight cryptography standards in 2019. It is an addition-rotation-XOR (ARX)-type block cipher. It supports keys of size 128, 192, and 256 bits. The number of rounds for these key lengths are 24, 28, and 32, respectively. Its block size is 128 bits.

Two dedicated keystream generators for lightweight stream ciphers, *Trivium* and *Enocoro*, are specified in ISO/IEC 29192-3:2012. Enocoro supports key sizes of 128 bits and 80 bits. Trivium supports only 80-bit keys. It should be noted that 80-bit keys can provide very short-term security and NIST expects at least 112-bit security from symmetric key encryption algorithms since 2019.

Note that if a blockchain is going to be implemented on resource constrained devices, a choice of a lightweight hash function might be better for performance and power consumption. Three lightweight hash functions that are suitable for applications or devices requiring lightweight cryptographic implementations are specified in ISO/IEC 29192-5:2016:

1. *Photon* supports message digests of 80, 128, 160, 224, and 256 bits.
2. *Spongent* supports message digests of 88, 128, 160, 224, and 256 bits.
3. *Lesamnta-LW* has a message digest of length 256 bits.

NIST initiated a lightweight authenticated encryption competition [41] in 2019 with optional hashing capabilities. Competition is aimed for constrained environments so that algorithms are expected to provide acceptable security and performance for these devices. The submission deadline for this competition was February 2019 and it received 57 candidate algorithms where 56 of them were accepted in April 2019 to participate in the first round. Thirty-two of these competitors were selected in August 2019 [13] for the second round. NIST organized three Lightweight Cryptography Workshops in July 2015, October 2016, and November 2019. The fourth workshop was held recently online in October 2020 due to the pandemic. On 29 March 2021, NIST announced 10 algorithms as the third-round candidates (finalists) as follows: *Xoodyak, TinyJambu, Sparkle, Romulus, Photon-Beetle, ISAP, Grain128-AEAD, GIFT-COFB, Elephant, and ASCON*. The winner(s) is expected to be announced and standardized around 2023.

13.8 Conclusion

Since the introduction of Bitcoin in 2008, cryptographic hash functions and digital signature algorithms have played an important role in the security of blockchains. We require preimage and second preimage resistant hash functions for blockchain security and near-preimage resistant hash functions for cryptographic puzzles. Broken hash functions like SHA-1, MD4, and MD5 should be avoided in blockchains. Moreover, lightweight hash functions might be beneficial for blockchains designed for IoT.

The choice of elliptic curve for elliptic curve digital signatures or Schnorr signatures is very important and using standard elliptic curves can prevent known attacks on elliptic curves. Moreover, randomness is vital for signature schemes and users must be sure that they have good randomness/entropy source. Randomness of IoT RNGs should be checked and designers should pay attention to error codes that can be generated by hardware RNGs. Instead of using an entropy source output directly as random numbers, they should be fed to cryptographically secure pseudorandom number generators.

Development of a large quantum computer would solve intractable problems like ECDLP, DLP, and IFP on a quantum computer in polynomial-time. Although current technology does not allow building such a quantum computer and we do not know if such a computer will ever be built, blockchains should consider including quantum-secure digital signature algorithms in their set of available cryptographic algorithms to be future-proof.

There are many different IoT devices with different capabilities and in blockchains that require encryption, a lightweight encryption algorithm should be chosen according to the needs of these devices that are going to be used in the blockchain network. Although there are ISO/IEC lightweight standards with 80-bit secret keys, secret keys should be at least 112 bits to resist exhaustive key search attacks.

Use of zero-knowledge protocols for anonymity in blockchains shows that cryptographic algorithms might find themselves in blockchains in the future to solve different problems.

References

[1] Nakamoto, S., *Bitcoin: A Peer-to-Peer Electronic Cash System*. https://bitcoin.org/bitcoin.pdf (2008).

[2] Wang, X., Y. L. Yin and H. Yu, "Finding collisions in the full sha-1." In Shoup, V., ed.: *Advances in Cryptology – CRYPTO 2005*, Berlin, Heidelberg, Springer (2005) 17–36.

[3] Leurent, G. and T. Peyrin, "SHA-1 is a Shambles: First Chosen-Prefix Collision on SHA-1 and Application to the PGP Web of Trust," In Capkun, S., Roesner, F., eds.: In 29th USENIX Security Symposium, USENIX Security 2020, August 12–14, 2020, USENIX Association 2020, pp. 1839–1856.

[4] Stevens, M., E. Bursztein, P. Karpman, A. Albertini and Y. Markov, "The first collision for full SHA-1". In Katz, J., Shacham, H., eds.: *Advances in Cryptology – CRYPTO 2017*, Cham, Springer International Publishing (2017) 570–596.

[5] Schnorr, C., "Efficient Signature Generation by Smart Cards." *Journal of Cryptology* 4(3) (1991): 161–174.

[6] Neven, G., N. P. Smart and B. Warinschi, "Hash Function Requirements for Schnorr Signatures." *Journal of Mathematical Cryptology* 3 (2009) (1): 69–87.

[7] Maxwell, G., A. Poelstra, Y. Seurin and P. Wuille, "Simple Schnorr Multi-Signatures with Applications to Bitcoin." *Designs, Codes and Cryptography* 87 (2019) (9): 2139–2164.

[8] Shor, P., "Algorithms for Quantum Computation: Discrete Logarithms and Factoring," In: Proceedings 35th Annual Symposium on Foundations of Computer Science, 1994, pp. 124–134.

[9] Alagic, G., J. Alperin-Sheri, D. Apon, D. Cooper, Q. Dang, J. Kelsey, Y. K. Liu, C. Miller, D. Moody, R. Peralta, R. Perlner, A. Robinson and D. Smith-Tone, *Status report on the second round of the NIST post-quantum cryptography standardization process* (2020). doi: 10.6028/NIST.IR.8309.https://nvlpubs.nist.gov/nistpubs/ir/2020/NIST.IR.8309.pdf

[10] Miers, I., C. Garman, M. Green and A. D. Rubin, "Zerocoin: Anonymous Distributed e-cash from Bitcoin," In: 2013 IEEE Symposium on Security and Privacy, 2013, pp. 397–411.

[11] Ben-Sasson, E., A. Chiesa, C. Garman, M. Green, I. Miers, E. Tromer and M. Virza, "Zerocash: Decentralized Anonymous Payments from Bitcoin," In: 2014 IEEE Symposium on Security and Privacy, 2014, pp. 459–474.

[12] Ben-Sasson, E., A. Chiesa, E. Tromer and M. Virza, "Succinct Non-Interactive Zero Knowledge for a Von Neumann Architecture," In Fu, K., Jung, J., eds.: Proceedings of the 23rd USENIX Security Symposium, San Diego, CA, USA, August 20–22, 2014, USENIX Association, 2014, pp. 781–796.

[13] Sonmez, M., K. McKay, C. Calik, Dong and L. Bassham, "Status report on the first round of the NIST lightweight cryptography standardization process." *NIST Internal Report NISTIR* 8268 (2019). doi: 10.6028/NIST.IR.8268

[14] Yaga, D., P. Mell, N. Roby and K. Scarfone, "Blockchain Technology Overview." *NIST* 8202 (2018). doi: 10.6028/NIST.IR.8202

[15] Jean J., *TikZ for cryptographers*. https://www.iacr.org/authors/tikz/ (2016).

[16] Dobbertin, H., A. Bosselaers and B. Preneel, "Ripemd-160: A Strengthened Version of RIPEMD" In Gollmann, D., ed.: *Fast Software Encryption*, Berlin, Heidelberg, Springer (1996) 71–82.

[17] Percival, C., "Stronger key derivation via sequential memory-hard functions". *Whitepaper* (2016), pp. 1–16. Published at https://www.tarsnap.com/scrypt/scrypt.pdf

[18] Johnson, D., A. Menezes and S. A. Vanstone, "The Elliptic Curve Digital Signature Algorithm (ECDSA)." *The International Journal of Information Security* 1 (2001) (1): 36–63.

[19] Pohlig, S. and M. Hellman, "An Improved Algorithm for Computing Logarithms over GF(p) and its Cryptographic Significance (Corresp.)." *IEEE Transactions on Information Theory* 24 (1978) (1): 106–110.

[20] Pollard, J. M., "Monte Carlo Methods for Index Computation (mod p)." *Mathematics of Computation* 32 (1978) (143): 918–924.

[21] von Oorschot, P. C. and M. J. Wiener, "Parallel Collision Search with Cryptanalytic Applications." *Journal of Cryptology* 12(1) (1999): 1–28.

[22] Menezes, A., T. Okamoto and S. Vanstone, "Reducing Elliptic Curve Logarithms to Logarithms in a Finite Field." *IEEE Transactions on Information Theory* 39 (1993) (5): 1639–1646.

[23] Frey, G. and H. G. Rück, "A Remark Concerning m-Divisibility and the Discrete Logarithm in the Divisor Class Group of Curves." *Mathematics of Computation* 62 (1994) (206): 865–874.

[24] Satoh, T. and K. Araki, "Fermat Quotients and the Polynomial Time Discrete Log Algorithm for Anomalous Elliptic Curves." *Commentarii mathematici Universitatis Sancti Pauli* 47 (1998): 81–92.

[25] Smart, N. P., "The Discrete Logarithm Problem on Elliptic Curves of Trace One." *Journal of Cryptology* 12 (1999) (3): 193–196.

[26] Semaev, I. A., "Evaluation of Discrete Logarithms in a Group of p-Torsion Points of an Elliptic Curve in Characteristic p." *Mathematics of Computation* 67 (1998) (221): 353–356.

[27] Gaudry, P., F. Hess and N. P. Smart, "Constructive and Destructive Facets of Weil Descent on Elliptic Curves." *Journal Cryptol* 15 (2002) (1): 19–46.

[28] Menezes, A. and M. Qu, "Analysis of the Weil descent attack of Gaudry, Hess and Smart". In Naccache, D., ed.: *Topics in Cryptology – CT-RSA 2001*, Berlin, Heidelberg, Springer (2001) 308–318.

[29] Schoof, R., "Elliptic Curves Over Finite Fields and the Computation of Square Roots Mod p." *Mathematics of Computation* 44 (1985) (170): 483–494.

[30] Schoof, R., "Counting Points on Elliptic Curves Over Finite Fields." *Journal de Theorie des Nombres de Bordeaux* 7 (1995) (1): 219–254.

[31] Koblitz, N., "Cm-curves with good cryptographic properties". In Feigenbaum, J., ed.: *Advances in Cryptology – CRYPTO '91*, Berlin, Heidelberg, Springer (1992) 279–287.

[32] Jao, D., S. D. Miller and R. Venkatesan, "Do all elliptic curves of the same order have the same diffculty of discrete log?" In Roy, B., ed.: *Advances in Cryptology – ASIACRYPT 2005*, Berlin, Heidelberg, Springer (2005) 21–40.

[33] FIPS 186-4, *Digital signature standard (DSS)* (2013). doi: 10.6028/NIST.FIPS.186-4

[34] Shamir, A., "How to Share a Secret." *Communication ACM* 22 (1979) (11): 612–613.

[35] Groth, J. and M. Kohlweiss, "One-out-of-many proofs: Or how to leak a secret and spend a coin" In Oswald, E., Fischlin, M., eds.: *Advances in Cryptology – EUROCRYPT 2015*, Berlin, Heidelberg, Springer (2015) 253–280.

[36] Bogdanov, A., L. R. Knudsen, G. Leander, C. Paar, A. Poschmann, M. J. B. Robshaw, Y. Seurin and C. Vikkelsoe, "Present: An ultra-lightweight block cipher" In Paillier, P., Verbauwhede, I., eds.: *Cryptographic Hardware and Embedded Systems – CHES 2007*, Berlin, Heidelberg, Springer (2007) 450–466.

[37] Shirai, T., K. Shibutani, T. Akishita, S. Moriai and T. Iwata, "The 128-bit blockcipher CLEFIA (extended abstract)". In Biryukov, A., ed.: *Fast Software Encryption*, Berlin, Heidelberg, Springer (2007) 181–195.

[38] Hong, D., J. K. Lee, D. C. Kim, D. Kwon, K. H. Ryu and D. G. Lee, "Lea: A 128-bit block cipher for fast encryption on common processors". In Kim, Y., Lee, H., Perrig, A., eds.: *Information Security Applications*, Cham, Springer International Publishing (2014) 3–27.

[39] Tezcan, C., "The improbable differential attack: Cryptanalysis of reduced round CLEFIA". In Gong, G., Gupta, K.C., eds.: *Progress in Cryptology – INDOCRYPT 2010*, Berlin, Heidelberg, Springer (2010) 197–209.

[40] Tezcan, C. and A. A. Selçuk, "Improved Improbable Differential Attacks on ISO Standard CLEFIA: Expansion Technique Revisited." *Information Processing Letters* 116(2) (2016): 136–143.

[41] McKay, K., L. Bassham, M. S. Turan and N. Mouha, "Report on Lightweight Cryptography." *NIST Internal Report NISTIR* 8114 (2017): 1–27. doi:10.6028/NIST.IR.8114

14

Access Control and Data Security of IoT Applications Using Blockchain Technology

Usha Divakarla[1] and K. Chandrasekaran[2]

[1]*Information Science and Engineering, NMAMIT,*
  Nitte, Karkala, India
[2]*Computer Science and Engineering, NIT-*
  Karnataka, Surathkal, Mangalore, Karnataka, India

CONTENTS

14.1 Introduction

The most important power in today's world is information and organizations battle to keep theirs and their customer's information as secure as possible. Machine learning has enabled computers to learn from large amounts of data and make predictions on these

DOI: 10.1201/9781003203957-17

data. Thus, the need for information is ever-rising. There are also companies that are responsible for just data collection and analyzing and then sell the results of their surveys. Advertising companies use this data to learn more about the individual for targeted marketing. The world is increasingly moving towards a decentralized network of applications, users and devices. Due to the various issues of security, privacy, single points of failures and monopoly of operation that may occur in centralized systems, decentralizing the system seems to solve most of these issues. Controlling access to systems and data in a centralized environment is usually done by storing a list of identities (or attributes and roles) in a centralized system, where matching and verification is done on the same system. The term "centralized" used in our context doesn't refer to a single computer system in operation, but rather to a single organization controlling the operations over the data. Distributed or decentralized control system is the solution to allow owners of data to have complete control of how their data is managed and distributed. Blockchain is a technology has been leading the front in this area of decentralization since its advent. It all began with the success of the first blockchain-based cryptocurrency, Bitcoin [1/37], which allowed users to transact in electronic cash between peers without the involvement of a trusted third party. Bitcoin paved way for methods to achieve trust in multi-party untrusted environments. It enabled this by achieving an immutable ledger through the use of distributed consensus algorithms. Ethereum was the first successful implementation of a blockchain that provides a built-in Turing-complete programming language used to create contracts allowing to define state transition functions [2/12]. This paved the way for an increasing number of blockchain-based applications. While Bitcoin's application was limited to just sending of electronic cash around to peers, Ethereum's blockchain implemented a distributed virtual machine called the Ethereum Virtual Macine (EVM). Any code recorded on the blockchain in the form of contracts, when executed, runs on the EVM.

The following paper aims to describe a study on the application of blockchain in the domain of access control focusing more on decentralized access control. The following section deals a description of the survey methodology and the research questions. The second section begins with an introduction of access control. This section deals with requirements of an access control system, current methods employed and limitations that need to be addressed. The next section describes the blockchain in detail, with its structure, types, and characteristics. The sections after this provide the answers to the three research questions formulated. Towards the conclusion of the paper, we also address the drawbacks of our survey and the efforts made to tackle these drawbacks.

14.2 Survey Methodology

The survey conducted covers most of the notable research work that has been performed in the recent years in the field of access control systems using blockchain. A total of 163 papers published between the years 2016–2018 (inclusive) were collected from various sources including IEEE Xplore, ACM, Scopus, and Springer. By applying our chosen inclusion and exclusion criteria, the set of papers was reduced to 55. Further, while conducting the study, 7 papers were removed from consideration because they did not specifically use blockchain for access control. Figure 14.1 lists the distribution of the papers under study. This study summarizes most of the recent work and is intended for

FIGURE 14.1
Source of papers found.

readers who wish to gain an insight on the many ways blockchain technology has been integrated to provide access control implementations.

Exclusion Criteria: Secondary papers like review or survey papers, case studies and book chapters, duplicate papers, papers not available in English

Inclusion Criteria:

- Papers that introduce any form of novelty over existing or other proposed papers
- Blockchain technology has to be used in some part of the proposed method for access control.
- Primary focus should be on the access control of data in- stead of being a subset focus a larger security perspective

To conduct the survey the following set of research questions (RQs) have been developed:

- *RQ1: What is the background demographic data of the various research works?*
- *RQ2: How has blockchain been used in the proposed access control schemes?*
- *RQ3: What are the directions for future research in access control and blockchain?*

To answer RQ1, the papers were analyzed for statistical information like year of publication, distribution of domains, country of publication, and type of paper published. Graphical representations have been provided for easy understanding of these results. Along with these, attempts have been made to reason the results obtained.

Leading the research in the chosen field is IoT, health care, and the cloud sectors. Papers under these fields have been further analyzed qualitatively to assess the proposed method, find out the specific use of blockchain, and limitations of the method if any. This section aims to provide a summary of the various methods that have been developed in recent years to integrate blockchain with existing access control systems or to create an entirely different system using blockchain to overcome the existing limitations of access control systems. Results of these studies have been provided as an answer to RQ2.

RQ3 is aimed at providing the readers with directions to lead to their studies in this field of access control. Blockchain is a leading technology and is being integrated into several fields to solve problems of trust, centralization and mutability of information. Based on the understanding of the works presented in the past three years, an attempt has been made to guide researchers in what better ways a blockchain can be used to aid access control.

14.3 Access Control

Access control refers to the controlling of access to specific resources or systems. In the world of computers, this mainly refers to the controlling of access to systems, files, codes, and other information stored in any computer device. The term *computer device* isn't just limited to a personal computer and can refer to any device in today's world that can store information and may or may not be connected to any network. In case of devices not connected to any network, the term physical access control is more applicable. In this paper, we refer to access control mainly to refer to technical (or logical) access control that refers to any technological (or algorithmic) restrictions that we impose on the way a particular resource is to be accessed.

Similarly, a secure access control system must implement the following requirements:

- **Authentication:** This step includes the identification stage also as described previously. Authentication is required to ensure that only legitimate users are allowed to access the system. Various day-to-day use of authentication can be seen in the usage of usernames and passwords, biometrics, physical keys, etc. Software authentication schemes generally rely on usernames and passwords or digital signatures.

- **Authorization:** This refers to giving access to the resource that has been requested for by a user after en-suring that the user is legitimate and meets the access requirements. This step is usually encoded in the form of scripts called as access control policies. These policies specify which user can access which all resources and/or vice versa, that is, which resource can be accessed by which all users.

- **Auditing:** A list of all access requests made, accepted, and denied are often kept track of to analyze later, especially in the event of a security breach. This process is called auditing.

While these are the basic requirements of access control models that current centralized systems implement, the upcoming distributed access control systems have a need to ensure the following features also:

- **Publicly Visible Auditing:** The requirement of auditing has to be extended further to ensure that all users must be able to access the audit logs in a free and fair manner. In more constricted systems, at least the owners of the resources must be enabled to freely see the list of users that have requested for access to their resources.

- **Owner Controlled Policies:** Users owning resources must be given complete and granular control over their resources. This means that owners must be allowed to specify the policy themselves as to who can access their resources as well as what actions can be performed on these resources.

- **Identity-Preserving:** While maintaining a public log of all access requests is important, it is also important to ensure that users' identities aren't revealed. The pattern of access of resources can reveal personal and confidential information about a person such as his preferences and hence care as to be taken to ensure anonymity.

A number of papers that we found in our study use the term "privacy-preserving" as a foremost requirement of access control. The term deals with allowing owners able to control the access to their data as well as protecting the confidentiality from unauthorized users. In a broader sense, the term also encompasses the ability for users to be view who all have accessed their data. At the same time, the identity of the user and the access requests to the data should not be revealed to the open public. When all the above requirements are fulfilled, a privacy-preserving access control system can be said to be truly built.

14.3.1 Access Control Methods

The previously specified features are often encoded into access control policies using different models known as access control models. These models vary in the way they are implemented and the degree of granularity and control they allow over the resources. Different models may use different variables and functions to implement the access control rules. Often all the access control methods are classified into *mandatory access control (MAC)* or *discretionary access control (DAC)*. In MAC, permissions are given on a level based hierarchy. Administrators create levels and assign users to certain levels while rules are specified in a manner that a user can access all resources that are not above his level. DAC, however, provides access by the identity of the individual user. Usually this is done by maintaining a list of authorized users for each resource. MAC is more suited for governmental use cases where classifications (labels) are assigned to each file system object and the subjects (users) and rules are assigned based on these classifications. DAC is more often used in a personal scenario allowing owners control over access policies of their resources.

Some of the common access control methods are as follows:

- **Role-Based Access Control (RBAC):** This type of model makes use of roles or privileges assigned to users to specify the rules of access control. RBAC can implement both DAC as well as MAC. This model is suited to medium-sized organizations that assign a smaller number of roles to its employees. One limitation of RBAC is that it doesn't allow for fine-grained access rules specification.

- **Attribute-Based Access Control (ABAC):** This model has been developed as a successor to the RBAC model that allows the use of additional attributes belonging to either the user, resource, or environment or a mix of all these attributes. Boolean operations are often used to specify rules using these attributes. ABAC allows for a higher degree of granularity, but at the same time, it is difficult to configure the access for specific users whilst keeping other users with similar attributes away.

- **Capability-Based Access Control (CapBAC):** Capability is defined as a communicable and unforgeable token of authority. In context of access control, this token represents a set of access rights that are provided to the user owning the token. This advantage of this type of model is that it provides easy delegation of rights by the transfer of the token.

- **Identity-Based Access Control:** This form of access control can be thought of as a specific implementation of the ABAC model where the attributes used are related to the identities of the users or the resources being accessed. Access is usually granted when users authenticate themselves, for example, using a password. This allows for a very fine-grained control over the policies allowed to specify rules per user based.

The previous models are just a few of the commonly used access control implementations we found in our study. Apart from these, several cryptographic methods are also used to achieve access control such as attribute-based encryption (ABE) [1–8], identity-based encryption (IBE) [9], and hierarchical identity-based encryption (HIBE) [10]. Often access control methods use these cryptographic algorithms to implement the previously discussed models. Our study has revealed that several blockchain-based methods employ the use of one more of these encryption mechanisms.

14.3.2 Challenges

Most of the previously described models are the commonly used architectures in most access control systems. Each have their own advantages and limitations and are applicable to at least one domain of security. However, there are still some challenges that need to be solved relating to the use of these models:

- **Need for Trust:** Most access control systems currently are controlled by a single centralized entity. This requires users to trust that the central entity doesn't act maliciously over the data nor share it to other unauthorized third parties. Additionally, even key distribution is done centrally, which requires a certain amount of trust in the key management authorities as well.
- **Sharing of data:** Concerns related to security of access control systems and data breaches prevent service providers from being open to sharing of data. Data sharing is important to enable researchers to collect and aggregate resources required for their work.
- **Sensitive data:** Nowadays a lot of sensitive data like health care related or the household data collected from IoT devices are being stored and processed over the cloud. Such data should be allowed controlled over by the owner of the data.

These challenges can be addressed by the blockchain technology offering benefits of immutability and transparency. Many of the works propose a method to implement one of the aforementioned access control models and using blockchain to overcome the previous limitations. The following section explains blockchain in more detail.

14.4 Blockchain

Blockchain is a tamper-resistant, distributed immutable ledger that has proven to be the technology to overcome the limitations of centralized applications and move towards a decentralized system. Bitcoin was one of the first successful practical application of the blockchain technology, attempting to develop a pure peer-to-peer electronic cash system that wouldn't require a trusted third party and also helps prevent the double-spending problem [11–14]. Following the success of Bitcoin, a number of cryptocurrency-based technologies followed including Ethereum, a blockchain implementation attempting to build a technology on which all state machine concepts may be built [15–17], and Ripple, introducing a new low latency consensus algorithm and several more implementations.

14.4.1 Structure of the Blockchain

Blockchain is simply a data structure offering several features such as immutability, decentralization, and pseudo-anonymity (users' identities are preserved but their identifiers are traceable). The structure of the blockchain is shown in Figure 14.2. This structure however isn't compulsorily followed by all blockchain implementations. Some implementations have additional information stored like state, code, etc.

This section dives deeper into the workings of the blockchain that facilitate these features.

1. *Blockchain Nodes*: Nodes of a blockchain refer the computing devices that store the blockchain and perform the functions critical to its working. However, not all nodes may perform the same functions. Nodes are often classified as *full nodes* and *lightweight nodes*. Full nodes store the entire blockchain and are used for the distributed consensus protocol to publish the new block. A full node that publishes the block is called as the publishing node or the mining node in the case of Bitcoin. A lightweight node however does not store the complete copy of the blockchain and it must send its transactions to a full node for it to be accepted into the system [18,19].

2. *Cryptographic Methods*: The working of the blockchain to provide a tamper-proof, secure, and authenticated storage is mainly attributed to the extensive use of cryptographic approaches in its design. Some of the cryptographic techniques used in this data structure include hashing, digital signatures, and asymmetric encryption.

Hashing is an one-way cryptographic function produces a fixed length output from any given input. Additionally, given a hashed output, it is not possible to figure out correctly the corresponding input and it is not feasible to obtain two input values hashing to the same output value [18]. Hashing is commonly used for verifying the integrity of the data

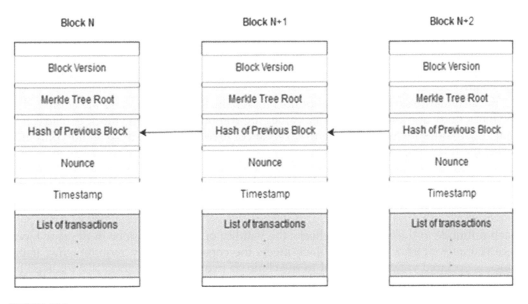

FIGURE 14.2
Blockchain structure.

stored on the blockchain. A common hashing algorithm called SHA-256 is used in Bitcoin. The popular cryptocurrency uses the complexity of the hashing algorithm to introduce a computational task that is expensive for nodes to easily create and publish blocks. This form of computation prevents attackers from spamming the network with invalid blocks or publishing requests [14].

Asymmetric cryptographic methods are used for two main purposes in the blockchain. The first is for signing transactions for maintaining integrity and authenticity. All transactions are signed by the issuer's private key. The second purpose is to generate addresses by hashing the user's public key. These addresses are assigned not only to users but also to smart contracts and are required to initiate a transaction with the user or the execution of a contract [18].

3. *Transactions and Blocks*: A transaction is the smallest unit in the blockchain. A transaction is an interaction or agreement between two participating nodes (or users) in the blockchain network. A transaction usually involves the transfer of some form of digital assets such as cryptocurrency tokens [18]. However, with the help of smart contracts [15], transactions can now be used to compute outputs on the network, transfer data, and even store some results on the blockchain. Smart contracts have now paved the way for transfer of data of any kind of structure such as authorization documents, currencies, business policies, or any other information that has to be stored securely and that cannot be tampered with.

A typical block in a blockchain consists of a block header and a block body. The header consists of fields such as block version, hash of the previous block, hash of all transactions in this block, timestamp, and nounce.

The second field is often built by storing the root node of the Merkle tree formed from the transactions [14]. The Merkle tree provides a fast and simple method to hash all the transactions and is used in Bitcoin [14] and Ethereum [15]. Ethereum however uses a modified Merkle Patricia Tree for its increased efficiency in insert and delete operations. Details like *timestamp*, *nounce,* and *block version* are implementation dependant and are used for the mining competition. All nodes compete to publish the next block in the blockchain for rewards given by the network. However, the blocks have to meet a certain criteria as specified by the used consensus model to be published onto the chain.

Transactions are sent to all nodes of the blockchain. These transactions are compiled in the block body and the block header values are computed. Once the block is ready, it is sent to all nodes for verification and acceptance. A block is treated as "valid" only if it meets the following two conditions:

• All transactions in the block must be valid.
• The block must conform with the chosen consensus policy.

Each full node in the blockchain checks the validity of the transactions in the new block received as well as whether the block meets the consensus policy requirements. If the block is declared valid, it is added to the chain. If two new valid blocks are received by the full node at the same time, the block that is longer (contains more transactions) and latest is added to the chain by default [18]. The following sub-section describes the consensus models of the blockchain in detail.

4. *Consensus Models:* In a distributed ledger like blockchain, it is important to decide which block will be published next, or rather which node will publish the next block in order to maintain consistency. A number of miner nodes will be trying to continuously gather transactions and create a valid block. However, the order and list of transactions collected will be different for each miner and a different block is generated by each miner node. When these blocks are broadcasted to the other nodes in the blockchain, different nodes may end up accepting different valid blocks. Therefore, there needs to be a method to select which block is accepted. Consensus algorithms help decide upon a condition which requires to be satisfied for a block to be accepted. Protocols like proof of work and proof of stake are two of the many algorithms that are in use in today's blockchains.

Proof of Work: In this model, a miner node gets to publish a block to the chain only by solving a computationally difficult task which is easy to verify. Bitcoin employs the proof-of-work (PoW) consensus model [14], which requires finding the hash value of the block that begins with a fixed number of zeros. The hash value of the block is changed by changing the nounce field in the header of the block. This problem is computationally expensive and it is believed that nodes that are willing to put in sufficient computing power to calculate the hash won't act maliciously. Once the nounce value has been found that produces the desired hash, the block along with the nounce is broadcasted to the network for verification. Moreover, being an expensive task, the chances of two valid blocks being generated at the same time is quite less. However, if two blocks are indeed broadcasted at the same time, then there exists a fork in the blockchain, that is, an inconsistency in the locally maintained blockchain between nodes. There are methods to resolve forks such as waiting for the next block to be published and then adopting the longest chain. Ethereum also currently uses the PoW consensus model, but is however shifting towards the proof-of-stake model for its light computational usage [15].

Proof of Stake: This consensus algorithm decides which user is going to publish the next block depending on the amount of "stake" the user has invested into the network. The amount of stake can be measured by several factors like age of the account, amount of cryptocurrency transacted till date or amount of cryptocurrency held, etc. The idea is that a user that has invested more time in the network has lesser incentives to subvert the system [18]. However, depending only upon the amount of cryptocurrency can gradually turn the system into a centralized system by making the richest user control the entire system. To prevent this, methods like randomization in BlackCoin and Nxt [20] and coin age-based selection in Peercoin [21] have been used. In contrast to the PoW model, PoS is less or barely computationally expensive.

There are several other consensus models developed by different blockchain implementations such as proof of authority, proof of weight, proof of elapsed time, proof of burn, and several more including modifications to PoW and PoS such as delayed proof of work and delegated proof of stake. However, the above two described consensus algorithms are the most popular ones in use.

5. *Smart Contracts:* Smart contracts have been termed since 1994 referring to computerized transaction protocols that execute the terms of a contract [18]. However, simple centralized programs cannot be used to create smart contracts as they are susceptible to attacks by malicious actors and error such as single point of failure. Also, centralized infrastructures require some amount of fees to be paid for carrying out these transactions. Blockchain technology has brought

with it the ability to carry out transactions in an untrusted environment. This feature of blockchain has been leveraged to introduce smart contracts in two of the most commonly used blockchains for decentralized applications, Ethereum [15] and the Hyper-Ledger Fabric [22].

Smart contracts (or simply known as contracts) can be described as lines of code that are executed on the blockchain (or participating full nodes) when predetermined conditions have been met and verified. They often are treated as another type of account on the blockchain having their own private storage, address, and account balance [15]. A contract is executed when a user sends a transaction to the address at which the contract is stored. Since the contract is stored on the blockchain, there is no way of changing the contents of the functions. All miner nodes of the blockchain execute the contract and take the contract from its current state to a new state. For the transaction to be valid, a majority of the nodes must agree upon the same two states. However, there are some blockchain implementations where all miner nodes do not execute the smart contract but instead just verify the results [18]. The execution of smart contracts also requires a small amount of processing fee to paid, which, in comparison to the conventional methods is far less. In Ethereum, this fee is paid in the form of "gas" [15].

With the advent of smart contracts, several applications of blockchain have emerged beyond cryptocurrencies such as electronic-voting, smart grid, smart properties, music rights management, etc. [23], leading to the second generation of blockchains. Among these blockchains, Ethereum's smart contracts are commonly used due the ability to write turing- complete code into the contracts that allow for more customization and advanced functions [23].

14.4.2 Permissionless vs Permissioned Blockchain

Permissionless blockchains allow anyone to enter the blockchain network and publish nodes. Any participant in the network can view and write onto the ledger. Hence, such types of blockchains are also called as public blockchains. Often PoW or PoS consensus protocols are used in this system to safeguard against bad actors in the network. Consensus mechanisms also include incentives to encourage good practices in the blockchain. The blockchains and their consensus algorithms are usually open sourced and available for anyone to use. As the blockchains are open to all, setup cost in a permission-less blockchain is often less. The popular blockchains in use, Bitcoin [14] and Ethereum [15], are examples of permission-less blockchains. Ethereum, however, has a permissioned blockchain implementation also for use.

Permissioned blockchains, on the other hand, require publishing users to be authorized by some kind of authority. The authority can often be a single organization (private blockchain) or a group of organizations (consortium blockchain). Since only a few nodes are allowed to verify the transactions, these blockchains are faster and offer higher scalability. Reading rights may be allowed to all participants or a selected few as decided again by the authority. The HyperLedger Fabric is an example of a permissioned blockchain [22].

14.4.3 Characteristics of Blockchains

Blockchain applications can now be found in several domains such as the Internet of Things (IoT), big data, cloud computing, edge computing, identity management, healthcare management, and even access control [24]. The following properties of blockchains make them suitable for integrating them into these fields:

- *Decentralization*: With the blockchain-based implementations, there is no longer a need to depend on a trusted third-party service provider to build or use the applications. The application is now self-sustaining depending on the users that are using it. Additionally, consensus algorithms help to maintain the consistency in this distributed network.

- *Transparency*: Data stored on the blockchain is stored on all nodes participating in the network. As such, the data stored is publicly visible for all users to see. However, cryptographic methods may be employed to ensure only some group of users can view the data while the remaining exist only to verify the integrity.

- *Integrity*: Data on the blockchain cannot be modified at any point of time. To perform a modification of the data requires changing the hash values of a long chain of blocks which is not feasible as the calculation of the hash of a single block itself is a computationally difficult task to perform. Hence, blockchain can be used to store information that can be used to verify the integrity of resources, files, or raw data. With use of sender's signatures and consensus protocols, the authenticity and consistency of stored information is further strengthened.

- *Anonymity*: Blockchain uses public keys to identify users on its network helping to preserve identity and maintain privacy. However, blockchain is often said to be pseudonymous as the public key used by a user is constant and a user's behavior can still be tracked if not the identity.

- *Auditability*: Since the data on the blockchain cannot be modified or erased once recorded, it can be used a immutable ledger of auditable information that can be used later for analysis.

14.5 Statistical Figures

This section provides the demographic information of the 48 papers that we have collected. The data answers questions like which publisher has published more blockchain and access control related papers, the trend of the papers over the years, whether the papers are of journal type or a conference proceeding, along with countries with which the first author or the institution of the authors is associated with and also the subdomain distribution of the papers (eg. cloud, healthcare, IoT). The data was collected for all the papers and formatted in a table. The data collected for the above questions has also been formatted into graphical figures for easy interpretation and understanding. Along with the figures, an attempt to find patterns in the data is also provided.

14.5.1 Year-Wise Distribution of Papers Published

As can be seen from the graph in Figure 14.3, we notice that the number of papers published in the fields related to access control and blockchain have almost tripled each year. The papers collected had been filtered to select only those published after 2016. This rise in publications could be mainly attributed to the rise of smart contracts in Ethereum,

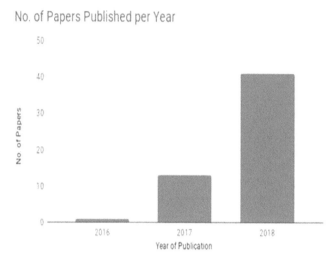

FIGURE 14.3
Papers published per year.

which was popularized in 2017. The ability to write Turing-complete code into the blockchain to program certain functions with every transaction, allowed several facilities like auditing and storage of custom records and structures. The provision of such services through an immutable ledger probably attracted many users to apply to various domains, one of them being blockchain. The work in securing access control with the help of blockchain is still an active field and we can expect even more papers in the future.

14.5.2 Types of Papers Published

We also analyze how many of the papers are published as represented by Figure 14.4 were of journal publications and how many were of conference proceedings. The general trend of papers is to find more conference proceedings than journal. However, the number of journal papers we have collected are quite considerable. This can be attributed

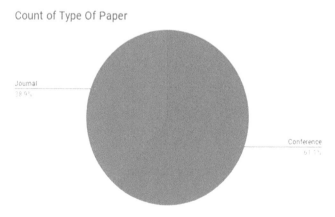

FIGURE 14.4
Journal vs conference papers frequency.

to the fact that blockchain is a relatively new topic and very few conferences exist that are solely blockchain specific. All of the works we have collected have been published in conferences related to *networks, IoT, security, communications,* etc.

It thus makes it necessary for authors to provide a detailed background on blockchain as well as access control before beginning their work description. The lack of blockchain-specific conferences may have also been one reason for a high number of journal papers.

14.5.3 Subdomain Distribution of Papers

The Internet of Things (IoT) is one of the fields attracting blockchain currently according to a recent mapping study conducted on the applications of the blockchain [25,26]. This can mainly be attributed to the decentralized and edge-resources nature of the block-chain. As IoT devices are constrained on their processing power and storage, they rely on higher layers for most of their tasks. These devices are also spread over large areas with limited communication power. A centralized system controlling the devices would lead to higher latency of communication. Additionally, data collected by these devices are sensitive and personal data of users which should be under the control of the owners themselves. To tackle such issues, a decentralized technology like blockchain has been used to provide privacy-preserving access control.

Other sectors, as shown in Figure 14.5, like cloud, storage, and big data all include research works related to storage of information in centralized or decentralized system with the inclusion of an access control system that is granular, user-driven, and allows sharing of data easily. The main problem with current storage solutions is the main-tenance of several independent data silos around the globe. There is a need to collect and aggregate necessary and authorized information among all these sectors leading to the use of consortium blockchain across storage providers to build a data sharing facility.

14.5.4 Distribution of Papers Based on Research Location

As Figure 14.6 shows, China is one of the locations that is focusing the most on research related to access control and the use of blockchains in access control, with more than twice the number of papers than any other country. This also complies with results from a

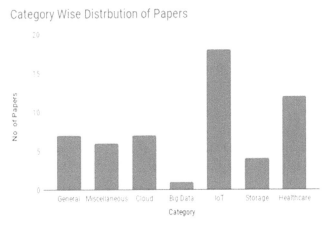

FIGURE 14.5
Category-wise distribution of papers.

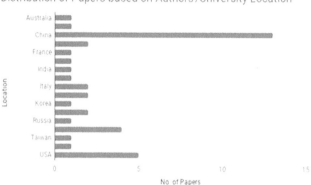

FIGURE 14.6
Location-based distribution of papers.

recent Systematic Mapping Study [25]. The next in line are universities in the USA focusing on blockchain-based research.

14.6 Blockchain-Based Access Control in Various Domains

14.6.1 Health Care

With the rise in benefits of data sharing and collaborative research, healthcare industries have shifted to the use of cloud-based storage of medical information, especially medical records in the form of electronic medical records (EMRS) or personal health information (PHI) as addressed by a few articles. PHIs differ from EMRs in the data they record. PHIs record the complete information about a patient, such as readings from their tests or smart devices. EMRs record data such as meetings with a doctor and the tests conducted. However, we will use the terms interchangeably here. These records often contain sensitive information about a patient; their age, physical, and mental characteristics; and health-related information. The most commonly used method of storage of such data is either in local storage servers or the cloud. However, there are problems associated with both methods. Storing of records locally restricts researchers from making use of the largely available medical data pool. Cloud storage risks revealing sensitive patients' data to third-party providers and losing control over how the data is managed or who can access the data.

1. *Challenges:* The following challenges are being faced currently in the healthcare industry, most of which have been attempted to be solved with the use of blockchain technology.

   - *Trust in Storage Providers*: EMRs are often stored in hospitals or external storage providers. Patients need to build trust in these parties before being willing to allow the storage on their EMRs. Even with proper access control mechanisms in place, issues can arise from within the controlling organization. Risks involve the unauthorized sharing of data for medical research or otherwise as well as changing of medical records. Due to this, patients often prefer to not digitize their medical records.

- *Historical Data*: The health insurance industry is one application where this facility is required. Insurance fraud happens when patients develop fake medical certificates to get insurance. This can happen easily as there is no recordkeeping facility for EMRs that can preserve patient's privacy as well as help provide a verifiable and immutable history of medical data to the authorized requesting agencies [27–30][P62].

- *Fragmented Storage*: Currently, EMRs are stored at the locations they are often generated or in separate cloud clusters. This limits researchers from being able to access the entire pool of available data [31–33][P52, P63].

A total of nine notable papers are shown in Table 14.1 that were found in the domain of access control in health care through blockchain application. The first paper to make a breakthrough in this domain is MedRec by Azaria et al. [31][P52]. We thus provide a detailed overview of the paper that formed the foundation of several healthcare-related papers that came after it. We also provide a detailed summary of other notable works in the domain of health care that are significantly different from MedRec and/or have solved any major issue arising in MedRec or other related works. Other shorter summaries are also provided to sum up the work done in this domain. MedRec is one of the first papers to be published that tackles the problems of local storage of EMRs. Different hospitals are unable to access a patient's medical data history due to the unavailability of sharing methods. This is usually done to protect the patients' privacy. MedRec proposes a blockchain based overlay over existing storage solutions to provide a access control system that allows patients to query their data themselves from anywhere and even share data with other medical institutions or researchers. This makes the design modular and easy to integrate with current solutions. Three different types of contracts have been in this framework. The first type is used to provide a mapping between a user and his Ethereum account. The second type links patients and medical record providers (medical institutions) and associated permissions allowing a patient to access their data via an available query string which can be applied on the provider's database to obtain the EMRs. The third contract provides a summary of all related medical data associated with any patient or a provider allowing easy access to fragmented data, especially for researchers.

The method proposed being one of the first, solves several issues relating to fragmented records storage, rules out centralized storage as an option, and provides a base for further systems to be developed using blockchain. However, the system does suffer from the following limitations:

- *Identity Exposure:* Ethereum account and user identification mappings are available on the blockchain for pubic viewing.

- *Records Storage:* Does not provide any information on how records can be securely stored. Assumes all health institutions use trusted local storage methods.

- *Publicly Accessible Policies:* Access permissions are stored openly on the public blockchain risking some kind of impersonation attacks.

- *Auditing:* Patients cannot view an immutable log of who all have accessed their medical records.

TABLE 14.1

Healthcare Study Papers

S.No	BC Type	Implemented On?	IP	Storage Type	Cryptography Methods	SC?	AR Audits	Owner Controlled	Customer Targeted	Blockchain Used For?
P52	Permissionless	Ethereum	Yes	Off-chain (cloud/local)	Public-key encryption	Yes	No	Yes	Researchers/Data Analysts and Hospitals	Storing access permissions on the blockchain
P19	Permissionless	Ethereum	Yes	Off-chain (cloud)	Public-Private Key	Yes	Yes	NA	Researchers	Auditing of access requests, sharing and transfer of data
P50	Permissioned	Hyper-Ledger	Yes	Off-chain	Public-Private Key	Yes	Yes	Yes	Researchers/Data Analysts and Hospitals	Storing of metadata of data stored off-chain and access policies
P61	Private & Consortium	JUICE	Yes	On-chain and Off-chain Indexing	Public Key Encryption with Keyword Search	Yes	Yes	Yes	Doctors	For data storage (private blockchain) as well as indexing of stored data (consortium blockchain)
P62	Consortium	Not Implemented	Yes	Off-chain	ABE+IBE for Medical Data, IBS for Digital Signature (C-AB/IB-ES)	Yes	No	Yes	Hospitals/Insurance Companies/Patients/Anyone wanting untampered records of whole medical data	Integrity and Traceability of Medical Data
P63	Public or Consortium	NM	Yes	On-chain Summaries and Off-chain Medical records	Public-Private Key	Yes*	NA	Yes	Patients/Researchers/Hospitals	Storing Summaries and Hash of Medical data/facilitate data sharing
P54	Permissioned & Private	Ethereum	Yes	On-Chain References and Off-chain Data	Symmetric, Public Key and Proxy Re-encryption	Yes	Yes	Yes	Patients/Doctors/Researchers	Storing query links to databases and hashed records to prevent altering. Also smart contracts for distribution
P48	Permissioned	Not Implemented	Yes	Off-chain data	Public-Private Key for User and Request Authentication, Verification and Generation	No	Yes	No (Policy isn't controlled by the user)	Patients, Authorized Researchers	Creating a log of access requests that have been made and granted

One of the major limitations of MedRec, which poses a slight threat to the privacy of the individuals by the revelation of the patient's identity is mainly due to the permission-less blockchain that is used by the framework. In a permission-less blockchain, any user may join the blockchain, access data and be able to verify the transactions. In a permissioned blockchain environment, users are given identities by a trusted identity provider. The identity provider helps to manage trust in the network by controlling access control within the network as well as the user's rights to validate new blocks. This kind of implementation has been deployed by Dubovitskaya et al. in [34][P50]. The authors have implemented an access control system over the HyperLedger permissioned blockchain that contains metadata as well as access control policies to control access to encrypted medical data stored in the cloud. These access control policies are defined by the patients who own the data themselves.

In a permissioned environment, users' identities are known and lesser chances exist for users to perform any malicious activities. The implementation on HyperLedger uses the PBFT consensus mechanism. Data is stored in two data stores, locally in the clinic associated with the medical record and another set of the organized patients' data in the cloud, encrypted with the patients' secret key. The framework ensures the following features, privacy, security, availability, and scalability. Privacy is provided by allowing a patient complete control over the access control policy over his/her data. Security attacks aren't very plausible since the network is permissioned and the identities of every user is known. With cloud-based storage and APIs, availability of data is also ensured. Scalability is ensured by the PBFT consensus protocol. However, the paper seems to target only one kind of medical record as mentioned and in future efforts, the authors will work on including more kinds of medical records, including images.

The next notable paper in our study of healthcare related blockchain research is [35] [P61] by Zhang and Lin. This paper is targeted more towards doctors being able to access historical healthcare related information of a given patient from all the healthcare institutions he/she has been in the past. The paper suggests that storing PHI in cloud-based services is insecure and poses security challenges which can be taken care of by storing them directly in blockchain.

The first blockchain is a private blockchain which is local to every health institution and stores records of patients who visit that institution. A private blockchain ensures faster transactions, lower cost, and greater security. The second blockchain is a consortium blockchain that stores secure searchable indexes of PHI in the respective private blockchain. These secure, searchable indexes are keyword search encrypted in a manner that fetches any doctor the results he/she would want without the revelation of the identity of the patient to any third-party listener. All identities and PHI records are encrypted using public-key encryption for security and privacy. The goals defined in the paper include data security and access control, privacy preservation, secure search, time controlled revocation (to ensure a given doctor can access the medical records only at the time he/she is authorized by the patient and not in future), and system availability. A consensus mechanism called proof of conformance has been suggested by the paper which checks for the valid authorization of the doctor while uploading the PHI to the private blockchain as well as checks if the secure search indexes are from a predefined set of indexes for system interoperability. Another method proposed in [27]P62 uses blockchain only for auditing and non-tampering of data. Access policies are signed by the patients and stored on the blockchain. Hospitals are responsible for encrypting the data using combined attribute-based/identity-based encryption and signature (C- AB/IB-ES) schemes and adding to the data pool controlled by the blockchain's consensus nodes. These nodes are responsible for verifying that the encryption meets the requirements of the policy as stated by the user as well as uploading of the data to the cloud and storing the address of the data on the blockchain. Any user wanting to query the data needs to search the blockchain for the address and then download from the cloud. Since the data is encrypted using identity and attribute based methods, only the authorized users will be able to decrypt the documents. As can be noticed, a blockchain is mainly used to ensure that the access control policy is maintained while encrypting the documents. All addresses of the documents are kept open on the blockchain for anyone to access.

[33] P63 describes MedBlock, another implementation of a integrity checking method using blockchain. The idea challenges the issue of medical records being fragmented in various medical institutions. Medical records are still stored in the hospital databases while only summaries and hash of the medical records are stored onto the blockchain. Medical records are encrypted using public key cryptography to ensure privacy of control over the data. Additionally, this paper contributes to a consensus mechanism that uses lesser power and is faster. Consensus takes place only among certain nodes, called leaders that are elected region wise. If any leader crashes, a re-election takes place. However, one limitation of this method is that it requires a trusted certificate authority responsible for the generation of public-private keys of the patients and the hospitals. The main idea of using blockchain to remove the need to trust is not met in this proposal. The proposal also employs a bread crumbs mechanism for efficient encrypted keyword searches through the ledger. Access control in this method is mainly ensured by the use of signatures.

Ancile is another framework suggested in [36] P54 that can be easily integrated into existing systems, thus offering interoperabilty and privacy. A permissioned blockchain is used to store hashes of references to medical records as well as the access control policies as set by the owner. However, smaller records can be stored directly on the blockchain. Proxy re-encryption techniques have been employed to enable the secure and privacy preserving transfer of such records stored. Proxy re-encryption also allows for the delegation of access rights allowing patients to let other users view their medical data. A major limitation of this method is the storage of medical records in the provider's database. Secure methods have to be ensured to secure not just the records but also the query links that are transferred over to users requesting for the data.

14.6.2 Cloud

Cloud computing is one of the most successful domains of computer science in the last decade, bringing everyone in the world closer to each other. Almost all data is now stored on the cloud for purposes of sharing, easy remote usage, and offering transparency of a centralized store. However, not everyone is aware of the risks associated the storage of data over the cloud.

A cloud is currently regarded as a trusted third-party storage solution provider. However, it may not be necessary that one's data stored over the cloud is as secure as is emphasized.

1. *Challenges:* A recent survey on access control issues in cloud computing mentions the following security issues in cloud computing:
   - *Loss of Data Control*: Once uploaded on the cloud, a user has no control over how the data is stored, secured or processed. Data storage and usage must be transparent to every user as if he alone is the sole controller of the data [37].

- *Need for Trust*: Cloud service providers (CSPs) do not provide any form of contract mentioning they will not misuse the data or for their own purposes. Additionally, they may decide to shut down their services any time, leading to a lack of trust in the CSPs. CSP may also leak data to third-party providers in anyonymized or non- anonymized form leading to breach of privacy. [38][P36]
- *Issues with Collaboration*: Data control misuse and leakage of data can also arise in Cloud Federations, where organizations share their data across cloud infrastructures to achieve a common goal. Not just data, even access control policies may be changed by any malicious participating party [39–42][P39/P47]
- *Access Control Key Distribution*: Another common issue with access control over the cloud is the distribution of encryption/decryption keys that are commonly used in several access control models. [43][P10]

2. *Current Research:* Out of the short-listed papers under our study, we found a total of six notable papers as shown in Table 14.2 under the cloud domain. Cloud is however, a domain that is a part of healthcare and even IoT. The six papers we consider here aim to tackle problems faced by users and cloud service providers (CSPs) in maintaining a secure, privacy-preserving environment for users' data. The papers under our study attempt to use a blockchain to suit the needs of a privacy-preserving cloud, having the facilities of data sharing, and auditing as well. At the same time, the aim is to ensure that the owners can control their data entirely even when stored on the cloud, including the various consumers who can access the data.

In [38]P36, the authors begin to describe the common issue of a cloud environment where the data owners have no information of how their data is being processed, stored or who is the data being accessed by. Data breaches and leakages to third-party data seekers is a major issue in such cases. The proposed solution involves combining existing cryptographic measures with blockchain's decentralized and tamper-proof solution to create a transparent, controlled, privacy-preserving environment also facilitating sharing of data among different data providers/seekers.

Data privacy is maintained by hierarchial identity based encryption (HIBE) and hierarchial identity based signature (HIBS) schemes. Data access, including the access policy is maintained entirely by the owner. The data owner themself generates a set of ID-based pair of keys to encrypt as well as sign the data intended to be shared with a service provider. Using per smart contract ID-based key, no link can be established between the smart contract and the data owner, thereby preventing any re-identification of data owners. For each service provider that the data owner intends to share the data with, a smart contract exists, defining the structure of data, access policies, as well as the actions that the service provider can perform over the data. Another form of smart contract also exists, when a second service provider/data seeker attempts to request for data from the first provider. A smart contract is published creating a transaction in the public blockchain thus notifying the owner, who goes ahead to create another contract with the second provider.

[40] P47 deals with issues faced by cloud federations (CFs). CFs are a collaborative program between cloud infrastructures wherein organizations share data in order to achieve a common goal. Issues relating to CFs include the leakage or misuse of data as well as unauthorized or illegal modification of access policy by a malicious user/participating entity.

TABLE 14.2

Cloud Papers Study

S.No	BC Type	Imple-mented On?	IP	Storage Type	Cryptography Methods	SC?	AR Audits?	Owner Controlled?	Customer Targetted	Blockchain Used For?
P36	Public or Con-sortium	NM	Yes	Off-chain	Hierarchial ID-based Public Key	Yes	Yes	Yes	General Cloud Customers	Transparent, Controlled Data Access and Sharing and Processing of data and Auditing
P39	Public	Ethereum	Yes	Off-chain data and on-chain access control policies and user at-tributes	Pedersen Commitment Scheme, Symmetric Key, Crypto Hash function	Yes	NM	Yes	Cloud Federations	To ensure users' identity attributes and access control policies cannot be modified
P43	Private	NM	Yes	Off-chain Data, On-chain decryp-tion keys	XOR based En-coding/Decoding Scheme	No	No	Yes	Users of ICN/ Content Providers in ICN	Storing the decryption keys for each user for the data provided by the content provider
P10	Public	Ethereum	NA	Off-chain Data	Public Key	Yes (Solidity)	Yes	Yes	Collaborators wanting to share cloud resources	Similar to voting system, owner's access control votes are stored on the contract in the blockchain. A user's contract also stores user's credentials and public keys for authentication and encryption
P9	Public	Ethereum	NM	Off-chain	CP-ABE	Yes	Yes	Yes	Users of untrusted cloud servers	To store access policies in a contract, link to the file and its hash for integrity checking
P42	Private (Imple-mented On)	Multi-chain and Timely CP-ABE	NA	Off-chain	Public Key (CP-ABE with Temporal Dimension)	Yes	Yes	Yes	Cloud users using data sharing or storing data for external use	For decentralization to avoid SPOF, to verify access control attributes and key management

The paper deploys an access control system similar to ABAC using users' indentity attributes. Intel SGX has also been used to protect the integrity and confidentiality of the policy enforcement process, which mainly uses a symmetric encryption approach. The user is able to reconstruct the key only if they satisfy the access control policy of the organization owning the data. The paper argues against the use of blockchain for enforcing the cryptographic policy for two reasons: the approach would be publicly visible to all and running the algorithm on the blockchain would be too costly. Blockchain is instead used to ensure that users' identity attributes or the defined access control policy cannot be easily modified by a malicious user but rather requires the verification of a certain majority of the participating organizations.

Information centric networking (ICN) deals with yet other issues related to cloud infrastructures wherein attackers can steal data from the various cache networks. Once cached at a location, data is no longer in the hands of the owner and cannot be controlled how it is treated or processed. To ensure that only the intended or authorized users obtain the data, authors of [44,45]P43 have discussed a private blockchain-based approach to distribute the data. While the data is encrypted by a single key, the decryption keys are stored on the blockchain for each user, encrypted by their public key so that only the authorized user can obtain the decryption key. This allows the content provider (CP) to not be online all the time.

The design of [44] P43 consists of splitting the content into n blocks and encoding into m (m¿n) blocks by the XOR encoding algorithm. Only n out of the m blocks can correctly decode the original data and only the authorized users can successfully obtain the right set of n blocks. The CP is responsible for encrypting the data, publishing the data to the blockchain as well as user registration along with the access policy thus giving the owner of the data in an ICN the complete control over his/her data. Upon registration with a CP, a user receives credentials which can be searched in the blockchain to obtain the decryption keys for that set of data.

Another issue similar to the one faced by cloud federations is the issue of shared ownership of data in the cloud. A single owner can delete files, revoke permissions and act maliciously without consulting the other collaborators. The paper [43] P10 proposes two options to deal with this issue, the second one being a blockchain based solution to reach consensus on access control decisions. Access control policies are stored on the blockchain via smart contracts. Access control decisions are taken by multiple owners and the votes are stored on the blockchain. Whenever a user requests access for a file, the blockchain verifies the request against the access control policy as well as the owners' votes stored on the blockchain.

14.6.3 Internet of Things (IoT)

The world is moving towards smaller and smarter devices that are handheld or integrated in the various daily-use applications such as refrigerators, TVs, watches, and many more. These small devices contain little processing power just sufficient to run their sensors or other hardware and perform data collection activities while the main brain making them "smart" usually lies in the cloud. A number of such devices around the world, continuously gather information from surroundings to build a large pool of data that is processed and studied on to make the devices more "smarter."

As the data collected is personal to a user, it must be up to the user to decide who can gain control over the data as opposed to the number of centralized systems these days who are in charge of the data once collected. These centralized systems make decisions on the data such as sharing them to third-party service providers in anonymized or un-anonymized form. Hence, the main aim of access control in IoT relates to giving every owner the complete control of privacy over their data. Table 14.3 shows the details of the papers collected representing access control in IoT. Data may be shared to third-party service providers; however, only after the permission has been granted by the user.

1. *Current Issues:*

   • *Need for Trust*: Centralized systems controlling the data from the IoT devices may sell to third-party service/application developers. Users are required to trust their storage providers. [46][P51]

   • *Control Bottleneck*: A large number of IoT devices may be connected to a single centralized server for access control decisions. This also acts as a performance bottleneck and suffers from single point of failure also [47][P58].

   • *Increased Latency*: Storage of data on a centralized cloud increases the distance between users and their data thereby increasing latency. This can be a limiting factor for several real-time applications that are dependent on the output of the sensors. [47–51][P45, P58].

   • *Complexity of Devices*: IoT devices have low storage and processing power making it difficult to directly integrate blockchain into these devices. The heterogeneity of IoT also poses an issue to integrating any system.

   • *Dynamic Nature of IoT Environment*: IoT devices are ever connecting and disconnecting to different networks. These devices can be added and removed as and when required into a network. Hence, there needs to be a method to dynamically define access control policies in the existing blockchain networks.

2. *Current Research:* One of the first attempts to provide access control methods in IoT using blockchain is *FairAccess*, proposed in [46][P51]. The main issue highlighted in this paper is the control of data by centralized access control providers. Sensitive data collected by the various sensors of IoT devices should be under the control of only the owner of the devices. The owners must decide how the data must be shared or distributed with third-party data seekers.

Being the first proposal in this cross-domain architecture, the authors also propose a new reference model for access control framework in the IoT perspective. They have also developed a implementation of a access control framework that is decentralized, user-driven, fair, transparent, and supports granular access control policies. The framework works by distributing authorization tokens which are analogous to crypto currency tokens used in Bitcoin. To gain access to a resource, a request transaction has to be initialized to a contract. The access policy defined in the contract validates the request and depending on the output, distributes the token or denies the request. The access policies have to be defined for each requestor-resource pair.

Other models making use of the authorization token approach as discussed in [46]P51 are the *emergence-based AC* model designed in [52]P6 and the *CapChain* model described in [53–55]P35. The model in [52]P6 also adds the use of reputation systems and reinforcement learning to extract the collective intelligence of a large number of IoT devices and allow the system to reconfigure themselves and their security policies as well as predict new or potential bad actors. In contrast to most implementations done on Ethereum, [53]

TABLE 14.3

IoT Papers Study

S.No	Blockchain Type	Implemented On?	Cryptography Methods	SC?	IP	AR Audits	Owner Controlled?	Blockchain Used For?
P51	Public/Private	Locally Implemented Blockchain	Yes (ECDSA)/Public-Private Key	Yes	Yes	Yes	Yes	Store, evaluate and enforce AC policy. Access tokens given out to enforce transaction based approach
P45	Public	Bitcoin testnet	Symmetric (AES-GCM)	No	Yes	Yes	Yes	Stores Access Permissions in the form of Transactions
P37	Public/Tested on Private	Ethereum (Private)	Public Key used for indentification	Yes	Yes	No	Yes	Stores list of all AC permissions to devices and resources
P40	NM	Ethereum	–	Yes	NM	Yes	No	Storing user-role and role-access rights relationships
P35	Public	Monero (Bitcoin variant)	Monero's Ring Confidential Transactions	No	Yes	Yes	Yes	Transaction based approach to AC
P41	NM	NM	NA	NM	Yes	Yes	Yes	Storing access policies, entity relationships, contextual data from sensors and auditing
P34	Public/Private	Ethereum (laptop + Raspberry Pi systems)	NM	Yes	NA	Yes	Yes	To store AC contracts and Judge Contracts (for penalty/rewards)
P1	Public & Private	NM	Public Key (for identification)	Yes	Yes	Yes	Yes	Stores Resource-Policy URLmappings
P22	Public/Private*	Not Implemented	NM	Yes	NM	NM	Yes	Storing and enforcing AC policies
P56	NM	Protocols implemented using C++, Blockchain no clue	Attribute Based Encryption (CP-ABE), ECDSA for Signatures to preserve identity	Yes	Yes	No	Yes	Storing AC Policies and IoT Device Management Data
P58	Implemented on Private Blockchain	Ethereum	–	Yes	Yes	Yes	Yes	Contracts containd Delegation Certificates with list of users with permissions
P60	Permissioned and Public	Not Implemented	Public Key	Yes	NM	Not Completely	Yes	Transaction based approach to AC
P6	NM	NM	–	Yes	Yes	Yes	Yes	Store AC Policies and Implement them. Similar to P51
P21	Private	Ethereum	ECDLP Based Public key	Yes	Yes	Yes	Yes	To Store Users' Access Policies for Immutability, which are verified by the IoT devices

P35 introduces CapChain built on the Monero blockchain making use of Monero's Ring Confidential Transactions to preserve privacy. The method makes use of the token approach to allow users to not just grant access rights but also delegate these rights to other users. The implementation models Cap- BAC by treating access rights as capabilities on the blockchain that can be transferred to users just like crypto currency. To pre serve identities, instead of using public addresses, transactions on the blockchain use derivable sub-addresses that function on the recipient and the delegator forming a hash chain as these permissions are delegated from user to user. The use of such sub-addresses allow for easy revocation of all rights from a user along with other users that may have derived their rights from this user. The implementation does suffer from drawbacks.

The limitation of this approach is that it doesn't provide the ability to revoke access easily as access revocation again requires a transaction on the chain. As the access policies are stored in the form of contracts on the blockchain, these are publicly visible which can form a threat [52][P6].

The authors of [46]P51 continue their work on FairAccess by using the second generation of blockchain to implement the architecture. This new generation of blockchain, called the account model, allows gained access control model through the use of advanced scripting language for smart contracts [56][P2].

To tackle the issue of distance between IoT devices and the cloud storage which results in an increase in latency, authors of [48]P45 have used a blockchain layer to provide an access control and data management framework over time-series IoT data stored at the edge of the network using a locality-aware decentralized storage system. Even cloud nodes can be used as storage nodes for this purpose. To secure the storage on edge devices, the IoT data streams are chunked, compressed, and encrypted at the application layer. Transactions are made on the blockchain to facilitate sharing of data with a service.

They include the datastream identifier and the public key of the service. Access requests are checked on the blockchain for a corresponding transaction by the owner and only then the key is shared with the requestor. This method suffers from the drawback of slow transaction speeds.

IoT devices constantly change managers or maybe are under the control of more than one manager at a time. However, manager devices are also low power devices that cannot easily execute the access control policies. Handling code consisting of managing access control, security, scalability, transparency, and concurrency will not be easy to execute on these manager devices. [57,58]P37 by Oscar Novo proposes an architecture using a single smart contract on the blockchain that is used to define all access control policies. The IoT devices are however not included as a part of the blockchain. A new entity defined as the management hub translates information between the blockchain and the devices. Managers are however considered lightweight nodes in the blockchain that are responsible for managing the access control permissions of the IoT devices under them by accessing the smart contract. A single contract automatically handles the issues of managing multiple devices and concurrency.

However, this architecture does not support the ability to audit the access control permissions. To obtain access permissions, the management hub nodes simply query the access control policy stored in the blockchain and translate the result of the request back to the devices. Transactions aren't supported as in current blockchain implementations; it takes a couple of seconds to minutes for the operation to complete. Additionally, modification of access control policy requires a transaction against the smart contract which might take some significant amount of time to reflect in the blockchain. This inherent limitation of the blockchain also poses as a threat to the strength of the access control model.

Another model proposed in [59]P41 involves the usage of multiple blockchain structures each for the following purposes:

- Storage of public credentials and relationships between entities (Relationships)
- Contextual data obtained from sensors (Contextual)
- History of permissions and denies for auditing (Accountability)
- Access control policies defined by the owners (Rules)

The novelty introduced in this paper involves defining relationships between devices or any other entities. Attributes and characteristics could be assigned to a pair of devices and could be used in the access control policy. The paper also suggests storing only information judged to be important in limited power devices, such as only storing rules associated with the device. Another option would be find another device with sufficient power and define a relationship in the relationships blockchain.

To overcome the limitations of processing in the IoT devices, [60–62]P34 opts for a blockchain-based architecture where the blockchain is stored on the IoT gateway nodes rather than the IoT devices. By including the blockchain on the gateway nodes, the problem of heterogeneity of the devices is also taken care of.

[60]P34 defines a model using multiple smart contracts types:

- Access control contracts which contain the policy for per subject-resource pair
- Judge contract to facilitate dynamic access rights validation by checking misbehaviour reports and assigning penalities if required
- Register contract that is used to define the above two types of contracts

Here also, IoT gateways act as agents for their local IoT devices and enforce the access control models for them. The paper introduces the ability to perform dynamic validation that can protect from attacks such as denial of service which can arise from multiple requests to access an unauthorized resource. However one limitation that the paper doesn't solve is that the interactions between the gateway and the IoT devices are assumed to be secure local interactions. IoT devices can be associated with a single gateway or sometimes be dynamic and link to any gateway; thus, a mechanism has to be developed to secure these interactions. Additionally, gateways are assumed to be trusted devices managing access control for their local IoT devices.

[63]P1 describes yet another model using blockchain and smart contracts for enforcing access control. However, the policies in this model are stored on the owner's local database while a URL is provided on the blockchain to this access policy. Any user requesting for the resource must lookup this policy and send to a defined smart contract along with his public key for authorization. An advantage of this system is that it is easy for a user to update their policy easily as it is available with them locally. The model also enables dynamic access policy creation whose addresses are stored on a private blockchain and are specific only to users who requested for them. However, the access policies defined on the public blockchain are visible to all which is one limitation of this model. Another limitation is that the access policies stored on the users' local database need to be made available at all times for access by the authorization contract.

[47]P58 uses the blockchain to store only capability tokens in the form of delegation certificates. The proposed method is developed to create an authorization delegation

mechanism in the distributed manner. User identification and authentication is performed on the cloud. The delegation according to the access policy is still performed by the owner on his/her device, but the various delegation relationships are stored on the blockchain for access authorization. Service providers upon receiving service requests check the blockchain for these delegations to grant or deny access. Revocation is also performed by invoking the smart contract holding these delegation certificates. This type of method overcomes the lack of scalability and revocation in current distributed CapBAC systems. However, the problem with this method of revocation is that the rights cannot be revoked immediately but require the transaction to be completed on the blockchain.

14.7 Future Directions for Research

Based on the study of the works on blockchain-based access control systems, we have attempted to provide an insight into what must be the grounds of future related work. This domain is fairly new and we hope that these insights form a foundation for future work to be structured around.

- **Auditing:** Blockchain technology has been made popular due to the collection of immutable records. Hence, auditing is a facility that is by default a feature of the blockchain. However, some implementations still do not currently provide this feature. Auditing of access control helps in understanding and improving the security system in place. Hence, an immutable ledger like blockchain must be deployed for its usage in auditing.

- **Blockchain as an Overlay:** Storage in blockchain is not easy as the size of the data stored in blockchains must be kept in control. However, size of data that needs to be monitored grows in size daily. Hence, off-chain storage methods are better opted for as compared to on-chain storage methods. Using the blockchain as an overlay over the existing storage method is the approach to go for in the future.

- **Transaction Speeds:** A number of transaction based approaches imitating the CapBAC model have been described, especially in the domains of IoT and cloud. However, these suffer from the slower transaction speeds in current generation blockchains. With the development of better consensus models and faster transaction speeds, these models can lay the foundation of future access control systems. However, currently, transaction speeds may still be made better with the use of permissioned blockchain [34][P50] or fewer consensus nodes [33][P63].

- **Easy setup and Interoperability:** Any system proposed has to be integrated with existing systems in little or no setup costs. Additionally, these systems have to be designed such that they can operate on any existing system. This requirement is necessary for consumers to be comfortable with the usage of a newer security system while the existing one is still performing as per needs. This concern also requires attention in the field of IoT containing high amounts of heterogeneity.

- **Dynamic Access Control Policies:** Most systems implemented offer the ability to define policies in the form of smart contracts that are statically controlled. However, access control also requires the ability to be able to dynamically control the stated policies and revoke access as and when required. Often the basic operations that must be considered by an access control system involve read, write, and no access. A user must be able to change these policies between one another at any time. This is another concern to be attented to in the domain of IoT.

14.8 Drawbacks of the Survey

This survey has been conducted by a thorough study of the selected papers and information has been collected in tabular form as well as summaries of proposed methods. However, the work may not be completely accurate and exhaustive due to the following drawbacks:

- A few notable papers may have been missed out during the paper screening and selection phase. Nonetheless, attempts have been made to include any notable papers that have been referenced significantly in our selected papers and may contribute to this research.
- The research in this field is fairly new and there isn't a lot of common structure to the issues tackled, solutions proposed, and limitations addressed. Due to this, a few fields in the tables provided are empty as the selected papers haven't addressed these concerns. To resolve this, we hope our future directions section will provide a structure for future researchers to address in their works.
- Language used to describe the methods may have varied from paper to paper, which could have led to different interpretations of the works carried out. We have attempted to understand thoroughly what the paper is trying to depict before writing our summaries, however, the reader is offered to read the actual article to understand the complete method employed. This drawback derives from the previous drawback and we once again hope that our paper sets the standards for terms and tokens to be used.

14.9 Conclusion

Even with the aforementioned drawbacks, this paper still provides a comprehensive overview of the work being carried out in access control using blockchain. A number of researchers are currently using blockchain for various security providing solutions. However, the research being carried out in this field, especially with the blockchain technology, is scattered around with different researchers using their own terminology and addressing only a partial set of issues and goals. Access control systems need to follow a certain standard to ensure easy interoperability and usage.

Not all methods proposed are perfect. Some suffer from the drawbacks of their own proposal and some from the inherent drawbacks of the blockchain technology. As the technology is still evolving, we can hope to find better research work in this domain in the future. Nonetheless, these works will still form a foundation and learning to create better access control systems using blockchains in the future. This paper may be used as a reference for future works to be structured and offers complete features that are currently lacking in most proposed systems.

References

[1] Bethencourt, J., A. Sahai and B. Waters, "Ciphertext-Policy Attribute-based Encryption," In 2007 IEEE Symposium on Security and Privacy (SP'07). IEEE, 2007, pp. 321–334.

[2] Bhaskaran, K., P. Ilfrich, D. Liffman, C. Vecchiola, P. Jayachandran, A. Kumar, F. Lim, K. Nandakumar, Z. Qin, V. Ramakrishna, et al., "Double-Blind Consent-Driven Data Sharing on Blockchain," In 2018 IEEE International Conference on Cloud Engineering (IC2E). IEEE, 2018, pp. 385–391.

[3] Goyal, V., O. Pandey, A. Sahai and B. Waters, "Attribute-based encryption for fine-grained ac- cess control of encrypted data," In Proceedings of the 13th ACM conference on Computer and communications security. ACM, 2006, pp. 89–98.

[4] He, Q., Y. Xu, Z. Liu, J. He, Y. Sun and R. Zhang, "A Privacy-Preserving Internet of Things Device Management Scheme Based on Blockchain." *International Journal of Distributed Sensor Networks* 14 (2018) (11): 1–11.

[5] Hu, S., L. Hou, G. Chen, J. Weng and J. Li, "Reputation-Based Distributed Knowledge Sharing System in Blockchain," In Proceedings of the 15th EAI International Conference on Mobile and Ubiquitous Systems: Computing, Networking and Services. ACM, 2018, pp. 476–481.

[6] Hussein, A. F., N. A. Kumar, G. Ramirez- Gonzalez, E. Abdulhay, J. M. R. S. Tavares and V. H. C. de Albuquerque, "A Medical Records Man- Aging and Securing Blockchain Based System Supported by a Genetic Algorithm and Discrete Wavelet Transform." *Cognitive Systems Research* 52 (2018): 1–11.

[7] Hwang, D. Y., J. Y. Choi and K.-H. Kim, "Dynamic Access Control Scheme for IoT Devices Using Blockchain," In 2018 International Conference on In- formation and Communication Technology Convergence (ICTC). IEEE, 2018, pp. 713–715.

[8] Jemel, M. and A. Serrhrouchni, "Decentralized Access Control Mechanism with Temporal Dimension based on Blockchain," In 2017 IEEE 14th International Conference on e-Business Engineering (ICEBE). IEEE, 2017, pp. 177–182.

[9] Boneh, D. and M. Franklin, "Identity-Based Encryption from the Weil Pairing," In Annual International Cryptology Conference. Springer, 2001, pp. 213–229.

[10] Gentry, C. and A. Silverberg, "Hierarchical id-based Cryptography," In International Conference on the Theory and Application of Cryptology and Information Security. Springer, 2002, pp. 548–566.

[11] Usman, W. C. "The Double Spending Problem and Cryptocurrencies." *Available at SSRN 3090174*, 2017.

[12] Chowdhury, M. J. M., A. Colman, M. A. Kabir, J. Han and P. Sarda, "Blockchain as a Notarization Service for Data Sharing With Personal Data Store," In 2018 17th IEEE International Conference On Trust, Security And Privacy In Computing And Communications/12th IEEE International Conference On Big Data Science And Engineering (TrustCom/BigDataSE). IEEE, 2018, pp. 1330–1335.

[13] Cruz, J. P., Y. Kaji and N. Yanai, "RBAC- SC: Role-Based Access Control Using Smart Contract." *IEEE Access* 6 (2018): 12240–12251.

[14] Nakamoto, S. et al., *Bitcoin: A peer-to-peer electronic cash system.* 2009. https://bitcoin.org/bitcoin.pdf

[15] Buterin, Vitalik et al., "Ethereum white paper: a next generation smart contract & decentralized application platform" *First version*, 2014.

[16] Cash, M. and M. Bassiouni, "Two-Tier Permission-ed And Permission-less Blockchain for Secure Data Sharing," In 2018 IEEE International Conference on Smart Cloud (SmartCloud). IEEE, 2018, pp. 138–144.

[17] Cha, S.-C., J.-F. Chen, C. Su and K.-H. Yeh, "A Blockchain Connected Gateway for Ble-based Devices in the Internet Of Things." *IEEE Access* 6 (2018): 24639–24649.

[18] Yaga, D., P. Mell, N. Rob and K. Scarfone, *Blockchain technology overview. Technical report, National Institute of Standards and Technology*, 2018.

[19] Yan, Z., G. Gan and K. Riad, "Bc-pds: Pro Tecting Privacy and Self-sovereignty through Blockchains for Openpds," In 2017 IEEE Symposium on Service- Oriented System Engineering (SOSE). IEEE, 2017, pp. 138–144.

[20] Vasin, P., *Blackcoin's proof-of-stake protocol v2. URL:* https://blackcoin.co/blackcoin-pos-protocol-v2-whitepaper.pdf, 71, 2014.

[21] King, S. and S. Nadal, "Ppcoin: Peer-to-peer crypto-currency with proof-of-stake." *self-published paper, August,* 19, 2012.

[22] Androulaki, E., A. Barger, V. Bortnikov, C. Cachin, K. Christidis, A. De Caro, D. Enyeart, C. Ferris, G. Laventman, Y. Manevich, et al., "Hyperledger Fabric: A Distributed Operating System for Permissioned Blockchains," In Proceedings of the Thirteenth EuroSys Conference. ACM, 2018, p. 30.

[23] Alharby, M. and A. van Moorsel, *Blockchain-based smart contracts: A systematic mapping study. arXiv preprint arXiv:1710.06372*, 2017.

[24] Gao, W., W. G. Hatcher and W Yu, "A Survey of Blockchain: Techniques, Applications, and Challenges," In 2018 27th International Conference on Computer Communication and Networks (ICCCN). IEEE, 2018, pp. 1–11.

[25] Bharadwaj, A. H. S., S. Dharanikota and K. Chandrasekaran, "Blockchain Research and Applications: A Systematic Mapping Study," In Proceedings of the 1st International Conference on Blockchain Technology. Springer, 2020.

[26] Sedgewick, P. E. and R. de Lemos, "Self- Adaptation Made Easy with Blockchains," In Proceedings of the 13th International Conference on Software Engi- neering for Adaptive and Self-Managing Systems. ACM, 2018, pp. 192–193.

[27] Wang, Hao and Yujiao Song, "Secure cloud-based ehr sys- tem using attribute-based cryptosystem and blockchain." *Journal of medical systems* 42 (2018) (8): 152.

[28] Wang, S., Y. Zhang and Y. Zhang, "A Blockchain-based Framework for Data Sharing With Fine-grained Access Control in Decentralized Storage Systems." *IEEE Access* 6 (2018): 38437–38450.

[29] Xia, Q., E. Sifah, A. Smahi, S. Amofa and X. Zhang, "Bbds: Blockchain-based Data Sharing for Electronic Medical Records In Cloud Environments." *Information* 8 (2017) (2): 44.

[30] Xia, Q., E. B. Sifah, K. O. Asamoah, J. Gao, X. Du and M. Guizani, "Medshare: Trust-Less Medical Data Sharing Among Cloud Service Providers via Blockchain." *IEEE Access* 5 (2017): 14757–14767.

[31] Azaria, A., A. Ekblaw, T. Vieira and A. Lippman, "Medrec: Using Blockchain for Medical Data Access and Permission Management," In 2016 2nd In- ternational Conference on Open and Big Data (OBD). IEEE, 2016, pp. 25–30.

[32] Banerjee, A. and K. P. Joshi, "Link before you Share: Managing Privacy Policies Through Blockchain," In 2017 IEEE International Conference on Big Data (Big Data). IEEE, 2017, pp. 4438–4447.

[33] Fan, K., S. Wang, Y. Ren, H. Li and Y. Yang, "Medblock: Efficient and Secure Medical Data Sharing via Blockchain." *Journal of medical systems* 42 (2018) (8): 136.

[34] Dubovitskaya, A., Z. Xu, S. Ryu, M. Schumacher and F. Wang, "Secure and trustable electronic medical records sharing using blockchain". In *AMIA Annual Symposium Proceedings*, volume 2017, page 650. American Medical Informatics Association, 2017.

[35] Zhang, Aiqing and Xiaodong Lin, "Towards Secure and Privacy-preserving Data Sharing in e-health Systems via Consortium Blockchain." *Journal of Medical* Systems 42 (2018) (8): 140.

[36] Dagher, G. G., J. Mohler and M. Milojkovic and Marella, P. B., "Ancile: Privacy-Preserving Framework for Access Control and Interoperability Of Electronic Health Records Using Blockchain Technology." *Sustainable cities and society* 39 (2018): 283–297.

[37] Charanya, R. and M. Aramudhan, "Survey on Access Control Issues in Cloud Computing," In 2016 International Con- ference on Emerging Trends in Engineering, Technology and Science (ICETETS). IEEE, 2016, pp. 1–4.

[38] Kaaniche, N. and M. Laurent, "A Blockchain- based Data Usage Auditing Architecture with Enhanced Privacy and Availability," In 2017 IEEE 16th International Symposium on Network Computing and Applications (NCA). IEEE, 2017, pp. 1–5.

[39] Alansari, S., F. Paci, A. Margheri and V. Sassone, "Privacy-Preserving Access Control in Cloud Federations," In 2017 IEEE 10th International Conference on Cloud Computing (CLOUD). IEEE, 2017, pp. 757– 760.

[40] Alansari, S., F. Paci and V. Sassone, "A Distributed Access Control System for Cloud Federations," In 2017 IEEE 37th International Conference on Dis- tributed Computing Systems (ICDCS). IEEE, 2017, pp. 2131–2136.

[41] Alessi, M, A. Camillo, E. Giangreco, M. Matera, S. Pino and D. Storelli, "Make Users Own Their Data: A Decentralized Personal Data Store Prototype Based on Ethereum and IPFS," In 2018 3rd International Conference on Smart and Sustainable Technologies (SpliTech). IEEE, 2018, pp. 1–7.

[42] Alexopoulos, N., S. M. Habib, and M. Mühlhaüser, "Towards Secure Distributed Trust Management on a Global Scale: An Analytical Approach for Applying Distributed Ledgers for Authorization in the IoT," In Proceedings of the 2018 Workshop on IoT Security and Privacy. ACM, 2018, pp. 49–54.

[43] Ritzdorf, H., C. Soriente, G. O. Karame, S. Marinovic, D. Gruber and S. Capkun, "Toward Shared Ownership in the Cloud." *IEEE Transactions on Information Forensics and Security* 13 (2018) (12): 3019–3034.

[44] Tan, X., C. Huang and L. Ji, "Access Control Scheme Based on Combination of Blockchain and Xor-Coding for ICN," In 2018 5th IEEE International Conference on Cyber Security and Cloud Computing (CSCloud)/2018 4th IEEE International Conference on Edge Computing and Scalable Cloud (EdgeCom). IEEE, 2018, pp. 160–165.

[45] Tapas, N., G. Merlino, and F. Longo, "Blockchain-Based IoT-Cloud Authorization and Delegation," In 2018 IEEE International Conference on Smart Computing (SMARTCOMP). IEEE, 2018, pp. 411–416.

[46] Ouaddah, Aafaf, Anas Abou Elkalam and Abdellah Ait Ouahman, "Fairaccess: A New Blockchain-based Access Control Framework for the Internet of Things." *Security and Communication Networks* 9 (2016) (18): 5943–5964.

[47] Xu, R., Y. Chen, E. Blasch and G. Chen, "Blendcac: A sMart Contract Enabled Decentralized Capability-based Access Control Mechanism for the IoT." *Computers* 7 (2018) (3): 39.

[48] Shafagh, H., L. Burkhalter, A. Hithnawi and S. Duquennoy, "Towards Blockchain-based Auditable Storage and Sharing of IoT Data," In Proceedings of the 2017 on Cloud Computing Security Workshop. ACM, 2017, pp. 45–50.

[49] Sharma, R. and S. Chakraborty, "B2vdm: Blockchain Based Vehicular Data Management," In 2018 International Conference on Advances in Computing, Communications and Informatics (ICACCI). IEEE, 2018, pp. 2337– 2343.

[50] Sifah, E. B., Q. Xia, K. O.- B. O. Agyekum, S. Amofa, J. Gao, R. Chen, H. Xia, J. C. Gee, X. Du and M. Guizani. "Chain-Based Big Data Access Control Infrastructure." *The Journal of Supercomputing* 74 (2018) (10): 4945–4964.

[51] Sukhodolskiy, I. and S. Zapechnikov, "A Blockchain-based Access Control System for Cloud Storage," In 2018 IEEE Conference of Russian Young Researchers in Electrical and Electronic Engineering (EIConRus), IEEE, 2018, pp. 1575–1578.

[52] Kalam, E., A. Abou, A. Outchakoucht and H. Es-Samaali, "Emergence-based Access Control: New Approach to Secure the Internet of Things," In Pro- ceedings of the 1st International Conference on Digital Tools & Uses Congress. ACM, 2018, p. 15.

[53] Le, T. and M. W. Mutka, "Capchain: A Privacy Preserving Access Control Framework based on Blockchain for Pervasive Environments," In 2018 IEEE International Conference on Smart Computing (SMARTCOMP). IEEE, 2018, pp. 57–64.

[54] Lin, C., D. He, X. Huang, K.-K. R. Choo and A. V. Vasilakos, "Bsein: A Blockchain-Based Secure Mutual Authentication with Fine- Grained Access Control System for Industry 4.0." *Journal of Network and Computer Applications* 116 (2018): 42–52.

[55] Lin, Di and Yu Tang, "Blockchain Consensus Based User Access Strategies in d2d Networks for Data-Intensive Applications." *IEEE Access* 6 (2018): 72683–72690.

[56] Ouaddah, Aafaf, A. A. E. Kalam and A. A. Ouahman, "Harnessing the Power of Blockchain Technology to Solve IoT Security & Privacy Issues." In *ICC* (2017): 1–7.

[57] Novo, Oscar, "Blockchain Meets IoT: An Architecture for Scalable Access Management in IoT." *IEEE Internet of Things Journal* 5 (2018) (2): 1184–1195.

[58] Novo, Oscar, "Scalable Access Management in IoT Using Blockchain: A Performance Evaluation." *IEEE Internet of Things Journal* 6 (2018) (3): 4694–4701. doi: 10.1109/JIOT.201 8.2879679

[59] Pinno, O. J. A., A. R. A. Grégio and L. C. E. De Bona. "Controlchain: Blockchain as a Central Enabler for Access Control Authorizations in the IoT," In GLOBECOM 2017–2017 IEEE Global Communications Conference. IEEE, 2017, pp. 1–6.

[60] Zhang, Y., S. Kasahara, Y. Shen, X. Jiang and J. Wan, "Smart contract-based access control for the internet of things." *IEEE Internet of Things Journal* 6 (2019) (2): 1594–1605. doi: 10.11 09/JIOT.2018.284770

[61] Zhu, Y., Y. Qin, G. Gan, Y. Shuai and W. C.-C. Chu, "Tbac: Transaction-based Access Control on Blockchain For Resource Sharing with Cryptographically Decentralized Authorization," In 2018 IEEE 42nd Annual Computer Software and Applications Conference (COMPSAC), volume 1. IEEE, 2018, pp. 535–544.

[62] Zhu, Y., Y. Qin, Z. Zhou, X. Song, G. Liu, and W. C.-C. Chu, "Digital Asset Management with Distributed Permission Over Blockchain and Attribute-based Access Control," In 2018 IEEE International Conference on Services Computing (SCC). IEEE, 2018, pp. 193–200.

[63] Dukkipati, C., Y. Zhang and L. C. Cheng, "Decentralized, Blockchain Based Access Control Framework for the Heterogeneous Internet of Things," In Proceedings of the Third ACM Workshop on Attribute- Based Access Control. ACM, 2018, pp. 61–69.

15

Lightweight Cryptography and Protocols for IoT Security

Runa Chatterjee

Department of Computer Science and Engineering,
Netaji Subhash Engineering College, Kolkata West
Bengal, India

CONTENTS

15.1 Introduction

Today, Internet of Things technologies help us to develop smart city services in a wide variety of ranges. In this domain, a large volume of sensors, variety of typologies in terminals, and various applications require proper security for maintaining sensitive information. Embedded systems are established in different domains like public sector, private sector, industries, complex infrastructures, wearable and mobile applications, etc. The Internet of Things enables us to make connections between people, objects, or animals over a network. It is performed without any human-to-machine or human-to-human interaction. Conventional cryptographic algorithms are generally used to save this sensitive data but they are not workable in a smart domain. In these pervasive environments, scarcity of resources led to the design of lightweight cryptographic primitives and protocols. Lightweight refers to small value of gate equivalents (GE), which are measured in terms of various aspects like key length, throughput rate, cycle rate, power consumption, and areas, etc.

Lightweight cryptography in the IoT is inevitable due to the following two reasons:

i. Efficiency of communication between end-to-end devices

To maintain end-to-end security, a symmetric key algorithm is used. In a resource constrained environment, battery consumption should be a minimum for the cryptographic operations. The lightweight symmetric key algorithm is applied to end devices for lower energy consumption.

ii. Increase applicability in lower resource devices

The lightweight cryptographic primitives require a smaller number of resources than the conventional cryptographic ones. This small amount of footprint increases the possibilities of more network connections within a small area.

The five primitives of cryptography are confidentiality, integrity, authentication, non-repudiation, and availability. Some of the cryptographic protocols include identification, authentication, grouping, ownership transfer, distance bounding, etc. But this paper focuses on lightweight primitives and protocols only.

Figure 15.1 shows lightweight cryptographic primitives. They are normally categorized into two parts. One part is for lightweight symmetric cipher and another one is for lightweight asymmetric cipher. Under asymmetric lightweight cryptography, RSA,

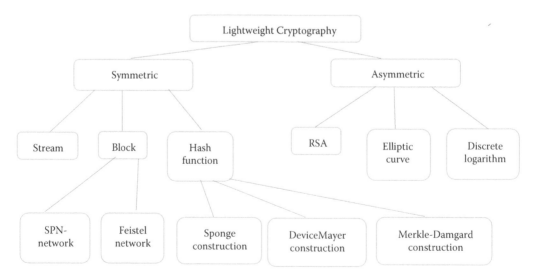

FIGURE 15.1
Variations of lightweight cryptographic primitives.

elliptic curve, and discrete logarithm are selected as subparts. At present, there is no favorable lightweight asymmetric cryptographic primitive that is comparable with conventional primitives. The detailed discussion over lightweight primitives has been discussed in [1]. Lightweight block cipher, stream cipher, and hash function all are under the symmetric key category, which are going to be discussed in Section 3, Section 4, and Section 5, respectively.

Lightweight block cipher – One of the symmetric lightweight cryptographic primitives is block cipher [2]. Based on the capability of devices, the implementation of block ciphers are categorized into three classes: lightweight, ultra-lightweight, and low cost. For lightweight implementations, a maximum of 3,000 logic gates can be allowed. Ultra-lightweight implementations are suitable in the resource constrained smart devices (in terms of memory requirement, power consumption, and computation capability). This type of implementation occupies up to 1,000 logic gates. A maximum of 2,000 logic gates are occupied by low-cost implementations. For software implementations, classification has been done on the basis of ROM and RAM requirements. More features and three types of block ciphers are briefly explained in Section 3

Lightweight stream cipher – Another symmetric lightweight cryptographic primitive is stream ciphers [3]. Here, the one-time pad (OTP) concept is used by the cipher known as Vernam cipher. The OTP generation process is based on fully random keystream. Each digit of the keystream is combined with one digit of plaintext to produce a digit of ciphertext. To make OTP unbreakable, the size of the keystream should be same length as plaintext. That's why it is considered impractical. The other features and three different stream ciphers are discussed in Section 4

Lightweight hash function – Every hash function has inputs that are referred to as messages, and the outputs are called message digests. The main property followed by all types of hash functions is it is impossible to recalculate input from output. That is why the hash function is called a one-way function. The attacker has no way of determining the content of the original message from the given the digest. The details are discussed in subsequent Section.

Lightweight protocol – The important factor of the IoT technology stack is IoT protocols. Hardware is meaningless in the absence of IoT standards and protocols. Such protocols authorize hardware to interchange data and are able to extract information from the end user. Lightweight data protocols are applicable to low-power IoT devices only. IoT network protocols are used for connecting devices. At the user side it interacts with hardware in the absence of the Internet. It's basically used as a wired connection for communication.

The rest of the paper is arranged as follows. Section 15.2 represents how lightweight block cipher, stream cipher, hash functions, and protocols provide security for IoT. Section 15.3 briefly explains the motivation of work in this field. Section 15.4 gives a brief description of three different lightweight block ciphers. In Section 15.5, the idea of a lightweight stream cipher has been given. Here, three popular stream ciphers have been discussed. The lightweight hash function working procedure and three different category hash functions have been discussed in Section 15.6. Section 15.7 includes three types of IoT security data protocols. Each primitive includes three different category ciphers or hash function and are discussed. In Section 15.8, within the discussion a comparative study is made on block ciphers, stream ciphers, and hash functions. Protocol comparison is done in a separate table. Finally, a conclusion has been drawn in Section 15.9.

15.2 LWC and Protocol for IoT Security

A huge number of nodes participating in an IoT domain is a major constraint. As devices have limited computational power, low memory, and worked in different operating environments, therefore, it is not possible to fulfill the security challenge in all aspects. The security requirement in the IoT is very much crucial due to three characteristics of IoT such as heterogeneity, dynamic environment, and resource constraint.

The key features of IoT security are as follows:

- Security in devices and data, authentication, confidentiality, and integrity of data.
- IoT scale-based implementation and running security operations.
- Entertain compliance requests and requirements.
- Use case dependent meeting performance requirements as per the use case.

The cryptographic primitives are generally classified as a block cipher, stream cipher, or a hash function. In the cryptographic environment block ciphers are treated as workhorses, which are discussed detail in [4]. A block cipher is comparatively better than a stream cipher in terms of implementation complexity and efficiency. It also has higher diffusion and error propagation rates than stream ciphers. Some of the specific designed block ciphers served as a purpose of security in terms of authentication or integrity protection. No single stream ciphers can do this.

Dynamic key generation techniques of lightweight stream cipher made cryptanalysis more tough by generating more random ciphertext. For this, various cipher primitives are updated for each input message to encrypt the next message. The lightweight stream cipher [5] maintains the perfect trade-off between performance and security in IoT devices. Stream ciphers are typically applied in those areas where simplicity and speed are mandatory requirements.

PHOTON [6,7], SPONGENT, and QUARK are the most newly brought out hash families that are based on sponge construction. It is an alternative approach to the classical Merkle-Damgård construction. Sponge functions utilize as operating mode is another attempt of improving compactness. As there is no feed-forward in sponge construction, it is possible to save a lot of memory registers which ultimately reduces the internal memory size. This is one of the constraints in IoT environment.

IoT data protocols are generally applied to connect low-power IoT devices. They provide communication between hardware and user without any internet connection. A specialized form of conventional network used in IoT domain. In this network, there are many things like gateways, sensors, which communicate to one another using IEEE 802.15.4 based lightweight communication protocols, such as MQTT and CoAP. XMPP is used for instant messaging application (like WhatsApp).

The detailed discussion regarding IoT security is discussed in [8]. In this paper, how lightweight block and stream cipher and lightweight hash functions are worked to make IoT more convenient to use are described in Section 3 How lightweight protocol maintains various security in different network levels are also discussed in the same section.

15.3 Motivation of Work

Today, our world is rapidly advancing towards the transformation of a smart and fast generation. In this smart world, a huge number of smart devices are connected to communicate between human to machine or human to human for faster transferring of data with security. But the constraints of a smart environment have reinforced another idea, the idea of lightweight cryptography. The National Institute of Standards and Technology (NIST) and International Organization for Standardization/International Electrotechnical Commission (ISO/IEC) standardized a number of methods for lightweight cryptography in an IoT domain. To maintain security in such a lightweight domain with all constraints in mind, various lightweight primitives and protocols have originated.

Under lightweight cryptography, the ISO/IEC 29192-5:2016 standardized PHOTON and SPONGENT as standard hashing methods, ISO/IEC 29192-2:2012 represent PRESENT and CLEFIA as standard block cipher method. The stream method Trivium was also selected by the ISO/IEC 29192-3:2012. In this paper, these standard methods have been discussed. The others are selected for their wide acceptancy in the industry. The three lightweight protocol selections are based on the strength of instant data transfer or variability of their industry usage or the popularity of multicast support.

15.4 Lightweight Block Cipher

One of the important lightweight cryptographic primitives is lightweight block cipher. Lightweight block ciphers belong to the symmetric key encryption category, which is operated on fixed length of bits. This primitive includes straightforward operations. Two types of networks are used to implement these types of ciphers. One is substitution and permutation networks (SPN) and the second one is the Feistel network. SPN uses substitution boxes (S-boxes) and permutation boxes (P-boxes) in each round for network implementation. Some mathematical operations together with the key convert a block of plaintext to a block of cipher text. The advantage of a Feistel cipher lies in its symmetric structure. Here, the encryption and decryption functions are almost the same except for the key scheduling method. Hence, to achieve low latency, the implementation code can be reduced by 50%. Some of the popular lightweight block ciphers are discussed below.

15.4.1 PRESENT

An ultra-lightweight block cipher, PRESENT [9] is based on a substitution permutation network (SPN) structure. It has 31 iteration rounds including block size of 64 bits. PRESENT supports 80-bit and 128-bit key variants. This cipher has three different layers in each round. Figure 15.2 displays the block diagram (including layers) of a PRESENT cipher.

Substitution layer: This first layer comprises 16 S-boxes where an individual S box has 4 bits of inputs and 4 bits of outputs; 16 parallel applications using 4-bit S-boxes have been done here.

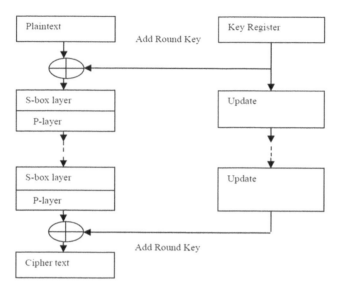

FIGURE 15.2
Block diagram of PRESENT block cipher.

Permutation layer: This is the second layer of PRESENT round. It performs 64 bits bit-wise permutation. It is a transposition technique that is simple and involves regular transformation of bits. Based on the permutation layer table, each bit is transferred to a fixed bit position.

Add Round Key: Key addition layer is the last layer of the PRESENT cipher. In this layer, the round key performs exclusive-OR operation with the 64-bit input text. It involves 61 bits left rotation of an 80-bit key register. Then the S-box is loaded with four important bits. Finally, least round counter bits perform XORed operation with the bits of K (K19 K18 K17 K16 K15). Thus, in each encryption process, the Add Round Key layer uses the updated key.

Three main things that make PRESENT faster and compact in hardware implementation are: 1) 4-bit S-box, 2) bit-wise permutation, and 3) exclusive OR (XOR) operation.

15.4.2 TEA

The tiny encryption algorithm (TEA) is a simple, safe, and fast block cipher as its name implies. It is a protected cipher and needs a minimum amount of storage space. TEA exhibits strong differential cryptanalysis. After a few rounds, it gives complete diffusion. The TEA uses 64-bit blocks as input and 128-bit keys. It performs 32 round operations. Figure 15.3 displays the working procedure of a TEA cipher:

    i. This cipher first divides the 64 bits of input data block into two 32-bit blocks. The left part and right part of the blocks are interchanged in their position in each round.

    ii. Next, a 128-bit key is broken into four subkeys of 32 bits each i.e., K[i] where $0 < = i < = 3$. It uses an integer delta constant 2654435769 ($2^{32}/$(golden ratio)). Here, a 128-bit key is taken as input. Multiples of delta are used in each iteration (mod 2^{32}) and addition operations are mod 2^{32}.

    iii. First, the right half performs a 4-bit left shift operation and then add with K[0].

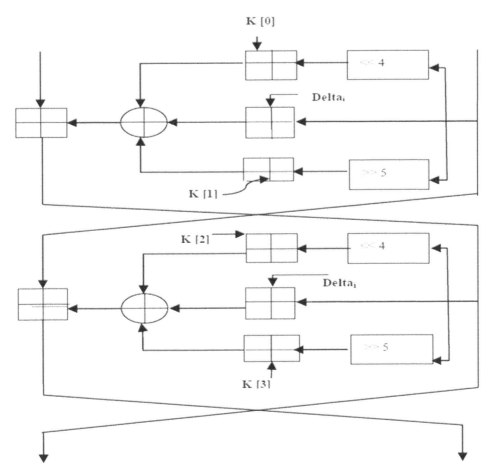

FIGURE 15.3
The block diagram of a single-round TEA cipher.

Again, the right half add with delta. The right half also performs 5 bits right shift operation and then add with K[1]. Then the output of these operations executes exclusive-OR operation and the resultant value is added to L (i–1). The result generates half of the block cipher which is used for the next iteration. Same set of operations are repeated for the next half round L [i–1] function.

15.4.3 CLEFIA

CLEFIA is an example of efficient and highly secure block cipher developed by SONY. It delivers advanced copyright protection and authentication. CLEFIA has a block size of 128 bits and key sizes of 128 bits, 192 bits, or 256 bits. Figure 15.4 describes the entire working procedure of a CLEFIA cipher. CLEFIA maintains a high-level security in implementation strategy of both world-leading hardware and software. It gives the highest hardware gate efficiency when it is implemented in hardware. It also gives faster performance on a wide variety of processors.

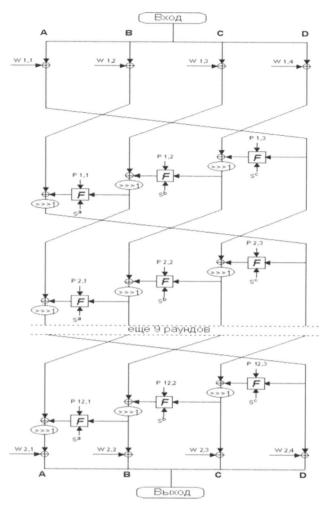

FIGURE 15.4
Block diagram of CLEFIA encryption.

15.5 Lightweight Stream Cipher

Another lightweight cryptographic primitive is a lightweight stream cipher. This type of cipher uses a key whose size is equivalent to the data. European Network of Excellence for Cryptology organized the eSTREAM competition, to find out new stream ciphers that might be suitable for widespread adoption. In stream cipher, the plaintext, an initialization vector and key stream together performs XOR operation to generate cipher text. The operation is performed in bit level using a key stream, generated from a key. Stream ciphers are potentially more compact for such a type of bit operation only. The hardware implementation of stream cipher is much simpler and faster than so-called block ciphers. In a stream cipher, two types of shift registers are generally used. One is a linear feedback shift register (LFSR), and the second one is a nonlinear feedback shift register (NLFSR). A stream cipher primitive is widely used in cell phones, wireless communication, etc. In this paper, three popular stream ciphers are discussed. They are GRAIN, TRIVIUM, and MICKEY.

The GRAIN family stream cipher is widely used and has greater implementation flexibility. GRAIN-128 AEAD version also supports authentication. TRIVIUM [10] is also popular but it supports only 80-bit keys. MICKEY [11,12] is less used compared to GRAIN and TRIVIUM. Due to irregular clocking, it gives less implementation flexibility.

15.5.1 GRAIN V1

GRAIN V1 [13,14] is one of the members of GRAIN family ciphers. There are three main building blocks of GRAIN, which are based upon a linear feedback shift register (LFSR), a nonlinear feedback shift register (NFSR), both are of k (80 or 128) bits, and a nonlinear filtering function. This cipher is initialized with the k-bit key K and the m-bit initialization value IV. The cipher output is keystream sequence of L bits. $(Ct)t = 0, ..., L-1$. Figure 15.5 illustrates the structure of GRAIN V1. The f1(m) and g1(m) functions definition are defined in equation 15.1 and 15.2, respectively. Ciphertext bit calculation is given in equation 15.3. The content of the LFSR is denoted by $Si, Si + 1, ..., Si + k - 1$. The content of the NSFR is denoted by $Bi, Bi + 1, ..., Bi + k - 1$.

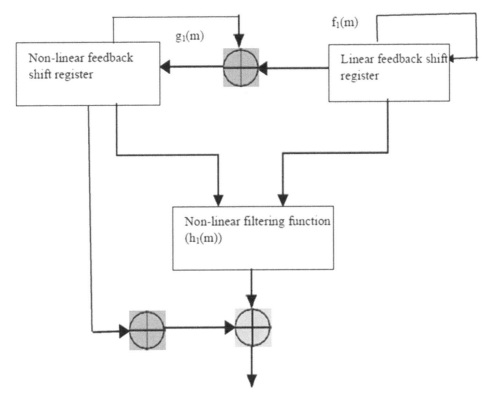

FIGURE 15.5
Block diagram of GRAIN cipher body.

For GRAIN v1, key size = 80 bits, m = 64. The LFSR feedback polynomial f1(m) is:

$$f1(m) = 1 + m^{18} + m^{29} + m^{42} + m^{57} + m^{67} + m^{80} \quad (15.1)$$

The feedback polynomial function of NFSR, g1(m) is defined as:

$$g1(m) = 1 + m^{18} + m^{20} + m^{28} + m^{35} + m^{43} + m^{47} + m^{52} + m^{59} + m^{66} + m^{71} + m^{80}$$
$$+ m^{17}m^{20} + m^{43}m^{47} + m^{65}m^{71} + m^{20}m^{28}m^{35} + m^{47}m^{52}m^{59} + m^{17}m^{35}m^{52}m^{71}$$
$$+ + m^{20}m^{28}m^{43}m^{47} + m^{17}m^{20}m^{59}m^{65} + m^{17}m^{20}m^{28}m^{35}m^{43} + m^{47}m^{52}m^{59}m^{65}m^{71}$$
$$+ m^{28}m^{35}m^{43}m^{47}m^{52}m^{59}: \quad (15.2)$$

The filter function of GRAIN V1 is h1(m0, …, m4) and is also the same as h0 (m0, …, m4), but the cipher output bit Ct^1 is derived as:

$$Ct^1 = \sum i \in A1\ Bi + 1 + h1(St + 3;\ St + 25;\ St + 46;\ St + 64;\ Bt + 63) \quad (15.3)$$

where A1 = f1{1, 2, 4, 10, 31, 43, 56}.

15.5.2 MICKEY

The full form of MICKEY' is mutual irregular clocking keystream generator. The original version of this cipher is MICKEY 1.0, which was at risk for statistical attacks. Later, the MICKEY 1.0 version was updated by MICKEY 2.0. The new variant reduces the gate equivalents (GEs) as well as power consumption without changing its security level. It is a lightweight stream cipher that can be applicable to both hardware and software implementation. Additional details can be found in [11].

The MICKEY 2.0 cipher includes two different registers for keystream generation. Figure 15.6 shows that one is linear register R and another one is nonlinear register S. These registers decide the value (1 or 0) for any given bit position within the keystream. The key and initialization vector (IV), used by MICKEY 2.0, both are of 80-bit. These key and IV are merged to initialize R and S registers where all the bit values are set to zeros.

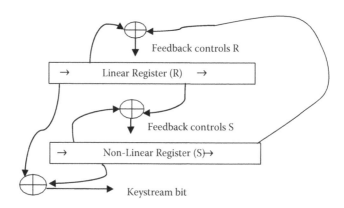

FIGURE 15.6
Key stream generation using linear and nonlinear register of MICKEY 2.0.

This is done to get started in the keystream generation process. Then, using the key/IV pair, they are loaded by calling the normal keystream clocking function.

Ideally, to hide any statistical patterns within the plaintext, the keystream pattern should be highly random and irreversible so that derivation of the key is quite impossible. Figure 15.6 illustrates the relationship between the linear and nonlinear register.

15.5.3 TRIVIUM

One of the finalists of eSTREAM Profile 2 is TRIVIUM, which is a standardized stream cipher for LWC (ISO/IEC 29192-3:2012). The designers of this cipher tried to simplify it as much as possible without compromising its flexibility, speed, or security. This stream cipher is bit-oriented and synchronous. Both initialization vector and secret key used by this cipher are of 80-bit. According to Figure 15.7, it is clearly visible that there are 288 bits internal state. Each state has three linear feedback shift registers (LFSRs) of variable lengths. It helps to skip building nonlinearity mechanisms for the keystream output. Around 2600 GE is needed for its implementation. The hardware implementation technology in standard CMOS includes 2017 GE. But for custom design implementation (using dynamic logic and C2MOS flip-flops), only 749 GE is needed. Although it was designed for hardware applications only, it performs satisfactorily in software also. The initialization process involves 2050 cycles, 55 cycles are required for setting the key, and

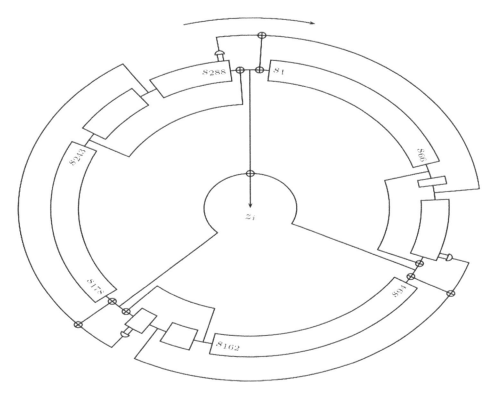

FIGURE 15.7
Structure of TRIVIUM stream cipher.

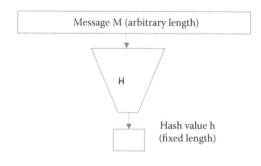

FIGURE 15.8
Illustration of a hash function.

12 cycles/bytes needed for the stream generation. Its simplicity makes it vulnerable to several attacks. A large volume of flip-flop utilization in hardware is the weakness of this cipher.

15.6 Lightweight Hash Function

In a cryptographic environment, a hash function generates a fixed-size output from a random length input. Figure 15.8 shows that technique. A cryptographic hash function is used to check the integrity and authenticity of information by creating a message digest.

There are two main parts in a hash function. One is construction and another one is compression functions. The compression function is iterated by the construction function. To make a hash function ideal, it should follow the random oracle property. Although this property does not exist, a security criterion must be followed by a hash function construction technique.

The three main security properties of cryptographic hash function (h) with a fixed length output are:

1. Collision-resistance: It is hard to find any two different messages x1 and x2 for a hash function h, such that h(x1) = h(x2).
2. Pre-image resistance: Reversing a hash function is computationally hard. It means it is difficult to find x from a hash value h(x).
3. Second pre-image resistance: If a hash function h produces hash value h(x1) as an output for input message x1, then it should be tough to find hash value h(x2) for any other input value x2 such that h(x2) = h(x1).

15.6.1 PHOTON

A small hardware-based lightweight hash function family PHOTON was designed by Guo et al. [7]. For internal key permutation, PHOTON includes an AES like primitive and a sponge- like construction. Sponge construction framework uses for very small internal memory utilization. But the squeezing process of sponge functions is very slow. Figure 15.9 displays the extended sponge framework. There it's extended this framework in such a way so that it gives interesting trade-off. This compact hash function requires 1120 GE. It

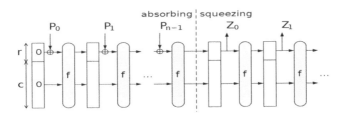

FIGURE 15.9
Extended sponge framework used by PHOTON family.

has 64-bit collision resistance security capability. PHOTON-n/r/r' representation signifies that its input and output bit rates are r and r', respectively. The output size n varies $64 \leq n \leq 256$. The size of the internal state depends on the size of the hash output: 100, 144, 196, 256, and 288 bits. The internal permutation P is applied to an internal state that contains d2 number of elements of b bits. This hash family uses 4-bit S-box of PRESENT and 8-bit S-box of AES. It has excellent area/throughput trade-offs.

15.6.2 SPONGENT

A lightweight hash function, SPONGENT was designed by Bogdanov [15]. It is based on a sponge-like construction and instantiated with the PRESENT-type permutations. A total of 13 variants are available for each collision/(second) preimage resistance level and implementation constraint.

The 4-bit substitution box (S-box) plays as a major block role for functional logic when considered in serial low-area implementation of this function. The design technique of SPONGENT includes bits permutation technique and S-box, which together fulfills the differential and linear properties.

The logic size of SPONGENT is close to the theoretically smallest size for its simple round function. In has the initial value, b-bit 0. For all variants of hash size n, the capacity is equal to either c or 2c. The message blocks perform XOR operation with the r rightmost bit positions of the state. The r bit forms parts of the hash output. Figure 15.10 displays sponge construction based SPONGENT function. Any linear approximation over the S-box (i.e., considers only one bit both in the input and output masks) is impartial. To make limitation of the linear hull effect discovered in round-reduced PRESENT this linear approximation is useful.

Figure 15.10: Sponge construction based on a b-bit permutation _b with capacity c bits and rate r bits. mi are r-bit message blocks. hi are parts of the hash value.

15.6.3 QUARK

Aumasson, in 2010, developed the first lightweight hash function QUARK [10]. QUARK hash function design technique is based on sponge construction [3]. A single security level is considered for reducing the memory requirement. This hash function permutation technique is based on two ciphers, one is block cipher KATAN and another one is stream ciphers GRAIN. Three different levels of security can be achieved by three different types of QUARKS.

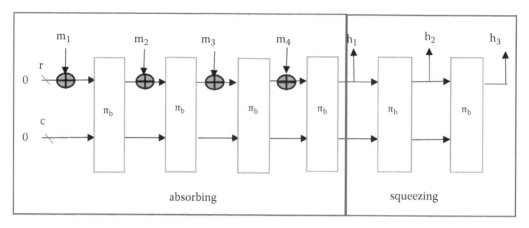

FIGURE 15.10
Basic sponge construction.

U-QUARK provides minimum 64-bit security against all types of attacks, D-QUARK gives 80-bit security, and T-QUARK gives 112-bit security.

For the moment, U-QUARK needs 1379 GE, D-QUARK requires 1702 GE, and T-QUARK requires 2296 GE. Three nonlinear Boolean functions f, g (same as GRAIN), and h are present for the internal permutation(P). All nonlinear Boolean functions are not the same for a single QUARK function. It includes one linear Boolean function p and involves P processes. The P processes are dependent on three phases. The phases are initialization, state update, and computation of output. Figure 15.11 shows the entire procedure of sponge construction of the QUARK function.

15.7 Lightweight Protocols

To handle resource constrained smart devices in an IoT environment, several computationally lightweight cryptographic protocols are used. IoT systems enable users to communicate with smart devices over the Internet. IoT-based systems require cryptographic primitives to encounter the security, privacy and trustworthiness. However, not all cryptographic primitives are applicable in an end-to-end manner in a resource limited IoT device. So, new security protocols are needed for effective use of cryptographic primitives in such a domain. The elaborate discussion of lightweight protocols (MQTT, CoAP, XMPP) is described in [16].

15.7.1 MQTT

One of the lightweight IoT data protocols is message queuing telemetry transport. In this protocol, simple data can flow from one device to another. Publisher-subscriber messaging model is its important feature. Its genetic makeup is basic and lightweight. The simpler architecture made this protocol in high demand in the market. It consumes low power for devices. It is also active on top of a TCP/IP protocol. To handle unreliable

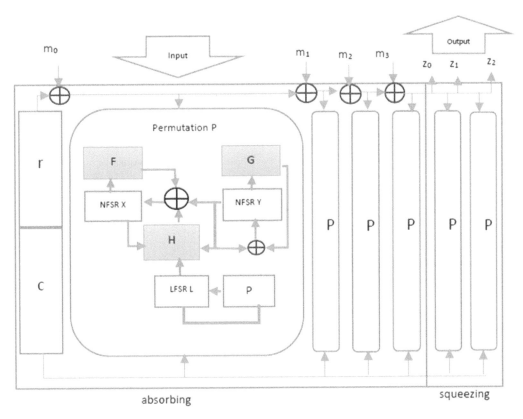

FIGURE 15.11
Sponge construction in QUARK.

communication networks, IoT data protocols were designed. Since the last few years IoT spreads its area vastly, this protocol was needed for handling large numbers of small, cheap, and lower-power objects in the network. Figure 15.12 describes the working procedure of a MQTT protocol.

In spite of MQTT's wide modification – especially on IoT-based industrial applications – specific data representation and device management are not supported by it. As a result, the implementation those is fully platform or vendor specific. Today, a MQTT protocol is used in various domains of industries, such as oil and gas, automotive, manufacturing, telecommunications, etc.

15.7.2 CoAP

An application layer protocol, CoAP (constrained application protocol) is treated as the base of data communication for the World Wide Web. It is a specialized web transfer protocol applicable to both constrained nodes and networks in IoT with low bandwidth and low availability. Internet Engineering Task Force designed the protocol. Although, the existing Internet structure is freely available and accessible by any IoT device. But a large volume of data usage makes it too heavy and consumes more power for most IoT

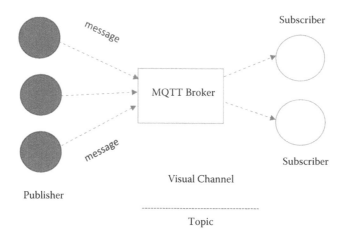

FIGURE 15.12
Working procedure of MQTT protocol (architecture).

applications. This has led to many within the IoT community dismissing HTTP as a protocol not suitable for IoT.

CoAp has mitigated such shortcomings by translating the HTTP model for usage in some restrictive devices and network environments. It has exceedingly low overheads. It includes multicast support and easily employable. These features make this protocol ideal for use in resource constrained devices such as IoT microcontrollers or WSN nodes. It's traditionally used for machine-to-machine (M2M) applications such as smart energy and building automation. Figure 15.13 describes the CoAP protocol working mechanism.

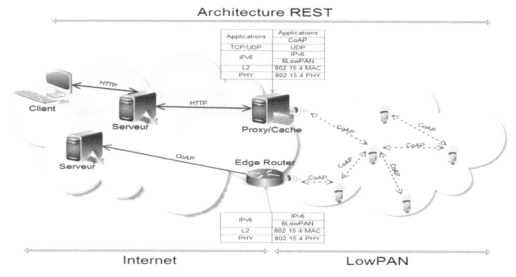

FIGURE 15.13
CoAP protocol.

15.7.3 XMPP

Extensible messaging and presence protocol (XMPP) is an open standard and it is ideal for near-real-time chat and instant messaging (like WhatsApp) by exchanging XML data over a network. Another name of this protocol is Jabber.

To store and organize data within documents a framework is used by extensible markup language (XML). For this, data interpretation in a wide variety of network endpoints are very easy irrespective of their hardware or software configuration. XMPP reliably sent XML data from one endpoint to another using a short snippet named "stanzas." For transmission XMPP uses the internet's transmission control protocol (TCP) through an intermediary server along the way. Figure 15.14 reflects the way of communication between end-to-end clients by passing stanzas.

In the case of instant messaging, XMPP and XML handle all most all the fundamental. It includes direct messages sending and receiving between users, status checking and communicating with users (presence) information, handling contact lists and chat list and, blocking communication to certain users.

15.8 Discussion

In this section a comparative study is discussed on three lightweight primitives and lightweight protocols. Under one primitive, three different ciphers are compared. They are compared with different parameters. Some comparisons are based on general parameters and some are hardware parameter. Table 15.1 illustrates the effective study on lightweight block ciphers. Table 15.2 and Table 15.3 reflect the comparative study on lightweight stream ciphers and lightweight hash functions. Lightweight protocols are compared in Table 15.4. Additional discussion on each cipher, function, and protocol are discussed in [7,16–20].

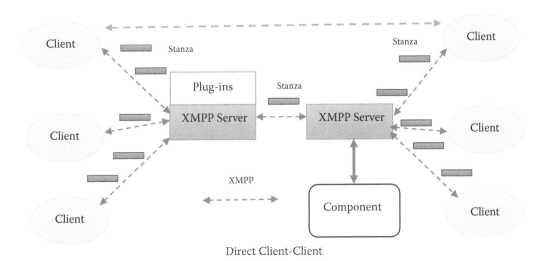

Direct Client-Client

FIGURE 15.14
XMPP protocol architecture.

TABLE 15.1

Comparative Study on Lightweight Block Ciphers

Light weight Block Ciphers	General Parameters				Hardware Parameters				Analysis
	Key Size (in bits)	Block Size (in bits)	Number of Rounds	Mode of network	Through put 100 KHz (Kbps) high	Latency (cycles/ block) lower	Power (microwatt) lower	Gate equivalent (GE) lower	Attack
PRESENT	80/128	64	31	SPN	206	31	2.20	2195	Side-channel attacks, related-key attack, improved differential fault analysis, Biclique attacks on full round, truncated differential attack
TEA	128	64	64	Feistel	6.25	512	7.00	2355	Equivalent keys attack
CLEFIA	128/192/256	128	18/22/26	Feistel	39	328	2.48	2488	Improbable differential attack on reduced round

TABLE 15.2

Comparative Study on Lightweight Stream Ciphers

Lightweight Hash Functions	Parameters				Strength	Weakness
	Key size (in bits)	Gate equivalent (GE)	Through put (Mbps)	Frequency (MHz)	Power (microwatt)		
GRAIN V1	80/128	1294/3239	177	177	7.772	(a) Bit-oriented stream cipher. (b) Throughput varies from 1 bit/clock to 16 bits/clock.	Possible attacks like algebraic, time, memory data trade-off, fault etc.
MICKEY 2.0	128	3188	250	250	8.701	Protects from any attack faster than exhaustive key search.	Heavily weak in related key attacks.
TRIVIUM	128	2580/4921	326	326	5.618	Well suited for the applications require a flexible hardware implementat ion.	(a) Correlations guess & determine attacks. (b) algebraic attacks (c) resynchroniza tion attacks in somewhere

TABLE 15.3

Comparative Hardware Analysis of Lightweight Hash Functions with Their Variant

Lightweight Hash Functions	Hardware Parameters (Operate under 100 kHz nominal frequency)							
	Hash size (in bits)	Gate equivalent (GE)	Power (microwatt)	Security		Latency (cycles/block)	Through put (kbps)	Strength
				Collision Resistance	Preimage Resistance			
PHOTON 128/16/16	128	1122/1707	2.29/3.45	64	112	156/1248	1.611 0.26	Provides strong security against linear and differential cryptanalysis.
PHOTON 160/36/36	160	1396/2117	2.74/4.35	801	124	180/900	2.70/20.00	
PHOTON 224/32/32	224	1736/2786	4.01/6.50	64	112	204/1716	1.86/15.69	
SPONGENT 128/128/8	128	1060/1687	2.20/3.58	64	120	70/2380	0.34/11.43	a. Improvement of Compactness is possible by combining SPONGENT hash with KTANTAN family. b. Provides simplicity in design with low area requirement.
SPONGENT 160/160/16	160	1329/2190	2.85/4.47	80	144	90/3960	0.40/17.78	
SPONGENT 224/224/16	224	1728/2903	3.73/5.97	112	208	120/7200	0.22/13.33	
U-QUARK	64	1379/2392	2.44/4.07	64	128	68/544	1.47/11.76	a. Secure against collision, second preimage, length extensions, and multi-collisions. b. Addition of decryption function can be made almost free by combining it with Piccolo cipher.
D-QUARK	80	1702/2819	3.1/4.76	80	160	88/704	2.27/18.18	
T-QUARK	112	2296	4.35/8.39	112	224	64/1024	3.13	

TABLE 15.4

Comparative Analysis of Lightweight Protocols

Lightweight Protocols	Type	Designed for	Standard	Transport	Qos	Security	Open Challenges and Efforts in Constrained Environments
MQTT	It is a client/server publish/subscribe messaging protocol. To provide ordered, lossless and bidirectional connections the protocol runs over TCP/IP	lightweight M2M communications	OASIS	TCP	3 levels	TLS/SSL	TLS version 1.3; MQTT-SN (based on UDP)
CoAP	It is a one-to-one protocol for transferring state information between client and server over the internet using UDP.	constrained devices	IETF (Internet Engineering Task Force)	UDP	Limited	DTLS (Datagram Transport Layer Security)	DTLS optimization
XMPP	It is a TCP communicat ions protocol based on XML. It's a distributed client/server-based architecture.	near-real-time exchange of structured data between two or more connected entities.	IETF	TCP	?	TLS/SSL	light-weight XMPP publish- subscribe scheme

15.9 Conclusion

In this paper, three lightweight primitives and three lightweight protocols in an IoT environment have been illustrated. Under lightweight primitives, how different block, stream ciphers, and hash functions played their role to maintain security in IoT have clearly been notified. In addition, hardware parameter analysis (e.g. throughput, area, latency, power, etc.) in a certain setup give an overall IoT security efficiency provided by different ciphers, hashes, and protocols. In the discussion section, a separate comparative study in a tabular form have been discussed. It gives a brief overview of the primitives and the protocols at a glance to the readers. This paper has written with the hope that it will help upcoming researchers by giving a consolidate idea of lightweight cryptographic primitives and protocols. It also helps researchers to improve security level from the basic knowledge of IoT security issues and requirements.

References

[1] Hammad, B. T., N. Jamil, M. E. Rusli, M. R. Z'aba and I. T. Ahmed, "Implementation of Lightweight Cryptographic Primitives." *Journal of Theoretical and Applied Information Technology* 95 (2017) (19): 5127–5133.

[2] Hatzivasilis, G., K. Fysarakis, I. Papaefstathiou and C. Manifavas, *A Review of Lightweight Block Ciphers*. Berlin Heidelberg: Springer-Verlag. 2017:141–184, doi: 10.1007/s13389-017-0160-y

[3] Manifavas, C., G. Hatzivasilis, K. Fysarakis and Y. Papaefstathiou, "A Survey of Lightweight Stream Ciphers for Embedded Systems." *Security and Communication Networks* 9 (2016): 1226–1246, doi: 10.1002/sec.1399

[4] Sehrawat, D. and N. S. Gill, "Lightweight Block Ciphers for IoT based Applications: A Review." *International Journal of Applied Engineering Research* 13 (2018) (5): 2258–2270.

[5] Noura, H., R. Couturier, C. Pham and A. Chehab, "Lightweight Stream Cipher Scheme for Resource-Constrained IoT Devices," International Conference on Wireless and Mobile Computing, Networking and Communications, Barcelona, Spain, Oct 2019.

[6] Meuser, T., L. Schmidt and A. Wiesmaier, "Comparing Lightweight Hash Functions – PHOTON & Quark," *Computer Science, Mathematics*, 2015.

[7] Guo, J., T. Peyrin and A. Poschmann, "The PHOTON family of lightweight hash functions". In: Rogaway, P., ed.: *Advances in Cryptology – CRYPTO 2011. Lecture Notes in Computer Science*, vol 6841. Berlin, Heidelberg, Springer. 10.1007/978-3-642-22792-9_13

[8] Oh, S. and Y. Kim, "Security Requirements Analysis for the IoT," 2017 International Conference on Platform Technology and Service (PlatCon), 2017, pp. 1–6, doi: 10.1109/PlatCon.2017.7883727

[9] Shah, I. N. M. and E. S. Bin Ismail, "Randomness Analysis on Lightweight Block Cipher, PRESENT." *Journal of Computer Science* 16 (2020): 1639–1647.

[10] Aumasson, J. P., L. Henzen, W. Meier and M. N. Plasencia, "QUARK: A Lightweight Hash." *Journal of Cryptology* 26 (2013): 313–339. doi: 10.1007/s00145-012-9125-6

[11] Alamer, A., B. Soh, and D. E. Brumbaugh, "MICKEY 2.0.85: A Secure and Lighter MICKEY 2.0 Cipher Variant with Improved Power Consumption for Smaller Devices in the IoT." *Symmetry* 12 (2020) (1): 32. doi: 10.3390/sym12010032

[12] Babbage, S. and M. Dodd, "The MICKEY stream Ciphers". *New Stream Cipher Designs*, LNCS 4986, Berlin, Heidelberg, Springer-Verlag, 2008, 191–209.

[13] Banik, S., "Some insights into differential cryptanalysis of grain v1". Information Security and Privacy/" ACISP 2014. *Lecture Notes in Computer Science*, vol 8544. Cham, Springer, doi: 10.1007/978-3-319-08344-5_3

[14] Zhang, H. and X. Wang, "Cryptanalysis of Stream Cipher Grain Family," *National Natural Science Foundation of China*, (NSFC Grant No.90604036) and 973 Project (No.2007CB807902).

[15] Bogdanov, A., M. Knežević, G. Leander, D. Toz, K. Varıcı and I. Verbauwhede, "spongent: A lightweight hash function". *Cryptographic Hardware and Embedded Systems – CHES 2011. Lecture Notes in Computer Science*, vol 6917. Berlin, Heidelberg, Springer, September 2011. 10.1007/978-3-642-23951-9_21

[16] Kayal, P. and H. Perros, "A Comparison of IoT Application Layer Protocols through a Smart Parking Implementation," 2017 20th Conference on Innovations in Clouds, Internet and Networks (ICIN), 2017, pp. 331–336. doi: 10.1109/ICIN.2017.7899436

[17] Hammad, B. T., N. Jamil, M. E. Rusli and M. R. Z'aba, "A Survey of Lightweight Cryptographic Hash Function." *Journal of Theoretical and Applied Information Technology* 95 (2017) (19): 811–812.

[18] Manjulata, A. K., "Survey on Lightweight Primitives and Protocols for RFID in Wireless Sensor Networks." *International Journal of Communication Networks and Information Security (IJCNIS)* 6 (2014) (1): 29–32.

[19] Dizdarević, J., F. Carpio, A. Jukan and X. Masip-Bruin, "A Survey of Communication Protocols for Internet of Things and Related Challenges of Fog and Cloud Computing Integration." *ACM Computing Surveys* 6 (2019): 36–57. 10.1145/3292674

[20] Diedrich, L., P. Jattke, L. Murati, M. Senker and A. Wiesmaier, "Comparison of Lightweight Stream Ciphers: MICKEY 2.0, WG-8, Grain and Trivium", Corpus ID: 55027354, 2016.

Index